Product Assurance
Principles

Product Assurance Principles

Integrating Design Assurance and Quality Assurance

Sponsored by
American Society for Quality Control

Eugene R. Carrubba

Ronald D. Gordon

McGraw-Hill Book Company
New York   St. Louis   San Francisco   Auckland
Bogotá   Hamburg   London   Madrid   Mexico
Milan   Montreal   New Delhi   Panama
Paris   São Paulo   Singapore
Sydney   Tokyo   Toronto

Library of Congress Cataloging-in-Publication Data

Carrubba, Eugene R.
    Product assurance principles.

    New ed. of: Assuring product integrity. 1975.
    Bibliography: p.
    Includes index.
    1. Design, Industrial. 2. Reliability (Engineering)
3. Quality assurance. I. Gordon, Ronald D.
II. Carrubba, Eugene R. Assuring product integrity.
III. Title.
TS171.4.C38   1988   620'.0042   87-16876
ISBN 0-07-010148-5

1234567890   DOC/DOC   89210987

ISBN 0-07-010148-5

*The editors for this book were Betty Sun and Rita T. Margolies, the
designer was Naomi Auerbach, and the production supervisor was
Dianne Walber. It was set in Century Schoolbook.*
*It was composed by the McGraw-Hill Book Company Professional & Reference
Division composition unit.*

Printed and bound by R. R. Donnelley & Sons Company.

The first edition of this book was published under the title
Assuring Product Integrity by Lexington Books, D. C. Heath and
Company, Lexington, Massachusetts, Toronto, London.

Contents

Preface

There exists a pervasive trend toward largeness of all human endeavor. An apparently natural result of this is that the products of our labor are either of inferior quality or prohibitively expensive. People involved in a complex activity appear to become preoccupied with the activity itself, losing sight of the *objective* of that activity. A prime example is afforded by any product of today's industry, whether it be a physical product or a service. The various aspects of product creation, realization, and customer support have been delegated to various groups of people. One of these groups of people is charged with the assurance of *product integrity*. This group of people is skilled in what we shall call the *assurance sciences*.

The assurance sciences are a collection of engineering disciplines which have as their common objective the attainment of product integrity. They attempt to fulfill a major function performed by the "master craftsman" in days gone by—to assure that the product is worthy of bearing the maker's name. As product and production have become more complex and sophisticated, the master craftsman has been effectively eliminated. His unified approach to assuring product integrity was lost. The specialized disciplines that emerged to replace him resulted from different needs recognized at different times. As a consequence, there has been a tendency for product integrity to suffer: workmanship becomes inferior, products fail in early life and are quite costly to maintain, aesthetic design overrides safety, and software faults cause system "crashes." Having experienced these problems, users are demanding better products with stronger guarantees. Federal, state, and local governments are legislating controls to protect the user. And producers are devoting more of their resources to product assurance. In adapting to this history, the various member branches of the assurance sciences have, individually, achieved a certain maturity. It is now time to devote serious attention to their collective, concerted application.

The purpose of this book is to show how the assurance sciences function to ensure product integrity. The two main product assurance areas (design assurance and quality assurance) are treated, with emphasis on their interrelationships and their primary relationships to product design, test, manufacture, and application. Since the contents will be of interest to people of varied backgrounds, we have intentionally written in lay terms, avoiding the jargon of the specialists. The book covers the background, management, implementation, and future of the assurance sciences in 14 chapters. The background chapters describe the evolution of the assurance sciences, define the included disciplines, and discuss their common goals and needs—leading to an examination of various approaches to effective unification of these disciplines. The theme of effective unification is carried to a practical "how to" level in the following chapters, which deal with management of the available assurance techniques in order to achieve sufficient product integrity at reasonable cost. The body of knowledge forming the essence of the assurance sciences is then described. The emphasis here is on proper approaches, with reference to (rather than repetition of) the necessary detailed technical/mathematical explanations. The final chapter looks to the future of the assurance sciences—the most recent changes, career possibilities, and our prediction of where the field is headed from here.

Most books on assurance-related subjects concentrate primarily on military assurance programs and the development of evidence that specified requirements are met. This book gives greater attention to commercial situations—and whatever the situation, attempts always to relate the struggle to achieve product integrity to financial risk, profit, and loss. We feel that this book is unique in treating the assurance sciences, in total, at a level that is understandable to the nonspecialist and, at the same time, in sufficient detail to be of practical use.

It is intended that this book will be useful to (1) professional industrial management; (2) practicing engineers, particularly those currently engaged in one of the branches of product assurance; and (3) students of engineering, manufacturing, and management technology.

We would like to thank the groups at McGraw-Hill Book Company and the American Society for Quality Control (ASQC) for their efforts in producing our book. Special thanks are also due to Tammy Griffin of the ASQC for efficiently coordinating our activities with McGraw-Hill. Finally, we want to thank our wives, Nancy and Janis, respectively, for their quiet forbearance during the preparation of this manuscript.

Eugene R. Carrubba
Ronald D. Gordon

1

History

*It is required that all barrels shall stand this
proof, and that none be so received which do
not.*   *War Department Regulation
July 5, 1798*

1.1 Craftsmen

What is product integrity? For now, let us just say that product integrity is the goodness of a product. In the next chapter, we define product integrity in more formal terms. As we proceed through this book, we present the techniques that form the assurance sciences. Furthermore, we explore the disciplines or functions that we collectively refer to as product assurance.

When did the concern for assuring product integrity begin? How did the assurance sciences evolve? Where did product assurance get its roots? As we shall see, these aspects of product integrity have a curious history.

Whether we recognize it or not, we have had the goal of achieving product integrity for a long time. Maybe it wasn't recognized as such, but nevertheless it was there. You had better believe that the ancient Romans were concerned when their swords broke during battle or their chariot wheels fell off. However, they didn't say, "We have a problem in product integrity!" Conversely, the early Egyptians didn't gloat and remark, "By golly, we have shown that we can assure product integrity," as they assembled the pyramids which today stand as one of the Seven Wonders of the World. Yet, these monuments of the ages are long-lasting evidence of the know-how and perseverance the Egyptians demonstrated in making durable products, and so on. One could go back through time and cite many such examples of product integrity, or lack of it.

In these early days, and even much later, approaches to product integrity were far from formal in nature. For the most part they were based on pride—either that of the worker or of the master. Sometimes the master's pride in having a good product was directly translated to his slaves—in the form of threats. This might be considered a unique approach by today's standards.

The first real concern for quality began to emerge in medieval times with the guilds (around the year 1000). These associations required that apprentices undergo a long training period, and that those striving to become master craftsmen provide ample evidence of their ability. These rules were directed in some degree to product integrity (quality). In one sense, these guilds were crude versions of today's assurance-oriented professional societies (e.g., American Society for Quality Control).

However, as products such as machinery became more advanced, there was a resistance on the part of the guilds, which were really geared for the self-interests of the master craftsmen. Although the master craftsman did much for product integrity in these early days, he also hindered it by resisting attempts to bring in new innovations. Furthermore, there were struggles both within and between guilds, as well as between guilds and outsiders.

Nevertheless, the master craftsman did much to foster the early advances of product assurance. It was a way for him to promote his craft. Although it is true he was involved with custom-made low-volume products, the craftsman generally made his products good in terms of quality and durability. Considering the tools, equipment, and materials he had to work with, one can certainly marvel at some of his accomplishments. However, his products suffered because of lack of standardization.

During these times, an informal view of product integrity obviously existed. Pride was the key. The result was products that were, for the most part, good. The craftsman did it all himself. He designed, built, inspected, and tested his product until he was satisfied with it. He made sure that it was worthy of bearing his name. He had an interest in the integrity of his product—even though he may not have looked at product integrity quite the way we do today. For the relatively primitive products of yesteryear, this was not a bad approach.

Then came some semblance of formalization to the problem of assuring product integrity. In France, during the making of weaponry around the early 1500s, instructions were given on important measurements, and some details were even provided concerning inspection. Some of the products of these times, such as locks and clocks, required the precision that only a highly skilled craftsman could achieve. Since only a skilled craftsman could put the necessary parts together, gaining precision was an art. This resulted in the creation of many one-of-a-kind products. Standardization, however, was starting to creep forward in the early 1700s for the making of armor and weapons.

In the late 1700s and early 1800s, with Eli Whitney and his musket, came the first serious attempts at product assurance as we recognize it now. The basic concept applied by Whitney was that of interchangeability—an important principle of American manufacturing and a concept vital to the first roots of the product assurance "tree." Special metal-working tools, which were virtually nonexistent, had to be developed. Materials properties had to be closely examined. The ability to make precision measurements had to be advanced. Documented requirements (e.g., dimensions and tolerances) that could be monitored from an assurance viewpoint had to be generated. It was at this time that the formal beginnings of the assurance sciences came to being, with the first measurements to assurance criteria.* It was here, too, that the rudimentary beginnings of what we know today as "quality control" got started. Whitney and his collaborators applied such quality control techniques as:

In-process gauging

Testing and inspection

Defect prevention

In-process control

Inspection standards

Quality and workmanship standards

In addition, the making of interchangeable parts in quantity laid the groundwork for acceptance sampling concepts. The assurance sciences had attained an embryonic stage.

1.2   Industrial Revolution

Around the same time (early 1800s), the factory system was developing in various industries in the United States. The degree of mecha-

---

* Implicit in "science" is the ability to measure, and to repeat the measurements and obtain the same result.

nization was light; therefore, the worker's skill was still the most important factor. Product quality was tied directly to the worker or his immediate superior, usually a foreman. However, as industry advanced during the Civil War period, so did the factory system. As more industries arose around the mid-1800s, workers left their jobs to seek better opportunities elsewhere. Newly arrived immigrants were being used to meet the growing labor demand. Long-term employees were hard to find; new untrained workers were prevalent. Product integrity suffered—mostly from lack of identification with the product.

As a way of bringing product quality under control, the responsibility for quality was wrestled from the worker/supervisor and given to an independent agent (inspector). He or she was to decide whether the product was satisfactory. This approach seemed to work to a certain degree.

However, as the industrial revolution began to gain momentum in the 1900s, mass production and automation complicated the achievement of product integrity. The speed of machinery (output) was replacing the skill of the craftsman (quality). With this growth, however, came specialization of the quality control function. One group specified quality, another set the standards, still another did the inspection and test, and on it went. Everyone was involved, but no one group was responsible; still product integrity suffered! To compound matters, the need for a larger volume of uniform precision parts became more obvious; and so did the need for precision measuring techniques and equipments.

Nevertheless, some order was being made out of all this chaos. Progress was being made in the management fields. The concepts of scientific management were being developed and put into practice. Quality control efforts started to become more coordinated and effective. Dr. Walter A. Shewhart of the Bell Telephone Laboratories started to apply statistical methods, such as the use of statistical quality control charts, to help control quality in the early 1920s. The application of these techniques to mass production industries, though slowly recognized (and finally "pushed" by the U.S. War Department with the outbreak of World War II), did much to enhance product quality. There was some semblance of control being achieved by quality control!

1.3 Age of Complexity

And so it went, with quality control being the standard bearer of product integrity. However, products began to become more intricate, complicated, and sophisticated—and so did the processes necessary to make them. It soon became apparent that it was insufficient to be concerned solely with the quality characteristics of a product.

Early in the 1950s, commercial airlines began to become seriously concerned with the problem of electron tube reliability. Successful studies were undertaken by an organization called Aeronautical Radio, Inc. (now ARINC Research Corporation) in an effort to improve the reliability of these devices. Around the same time, December 1950, the Air Force created an Ad Hoc Group on Reliability of Electronic Equipment. Its mission was to study the overall reliability situation and provide recommendations that would enhance the reliability of equipment, while at the same time reducing maintenance. Then in late 1952, the Department of Defense established the Advisory Group on Reliability of Electronic Equipment. AGREE, as it was called, published a significant report in June 1957 addressing such important topics as reliability demonstration test requirements. With the advent of the missile age characterized by "one-shot" no-maintenance requirements, interest in reliability became more intense. More scientific approaches began to be applied to improve reliability—statistical analysis techniques, life testing, burn-in testing, design redundancy, etc. Products, both military and consumer, began to be assessed on their ability to operate successfully in the manner and for the purpose intended over a given time period, when used in a specified environment.* Furthermore, they failed less frequently and their life was being prolonged. Reliability had come of age—with a tremendous boost from the military.

However, reliability was not a panacea to product integrity. A 100 percent reliable product is an unattainable goal. When a complex product did fail, it was often difficult and costly to repair. The military became increasingly aware of the maintainability problem around 1954, particularly when they found out that the maintenance costs for their complex systems were approximately one-third of the total operating costs and that one-third of their total personnel were involved in maintenance and support functions. Consumers found they had similar problems. They found that it cost more to fix many products than to throw them away and buy new ones—if they could get them repaired at all. Products or equipment that were "down" were costly in both dollars and aggravation. The consumer could not cut the grass if the lawnmower couldn't be repaired; the flier at war could not fly if spare parts were unavailable; and so on. Thus sprang forth another product assurance discipline—maintainability—to meet the challenge. Maintainability began to become an integral part of the systems analysis and design process.

---

* This is one of the definitions often applied to reliability.

With the start of World War II, technological advances increased, resulting in a variety of new and complex products for both military and consumer use. However, it didn't take long to discover that many of these products were not properly designed for use by typical human beings. Operators made too many errors in the operation of radar equipment; accidents were attributed to human mistakes caused by design deficiencies. The human aspect was also an important factor in maintainability and maintenance. For example, if something failed in a piece of equipment and it was not within reach or outside the capabilities of a human being, it would be difficult to fix. Therefore, consideration was given to the user and repairer through still another new discipline of product integrity—human factors.

Still this wasn't enough to solve the problem of achieving product integrity. Who wants to use a product that is unsafe? You think you have a shock-free drill; you stand on a damp surface and begin to drill—and zap! Once again there was concern on the part of the military—they began to demand equipment that was safe for the operator, was safe for the repairer, and caused no damage to other parts of the equipment. Similarly, the general buying public, thanks to activists like Ralph Nader, began to get the message in the consumer product area. Consumer liability suits are now commonplace items in the case of accident, damage, or death due to faulty products. With these concerns came yet another discipline called, oddly enough, product safety.

The computer age spawned a new generation of product integrity problems and challenges. Until that time, the focus of product integrity concerns was on hardware. However, the horizon expanded with the use of computers—the dimension of software and the interaction of the software with hardware were added. Although the importance of computers was understood early in their history, particularly during World War II, it was not until the 1960s that their use became more common. As computer use became more extensive, so too did the recognition of computer-related problems. System outages due to software "bugs" struck a serious note, which was amplified by software maintenance inadequacies. Finally, in the late 1970s, the military began a drive for software quality standards and methodology to assure software quality. For the first time in its history, when related to software, the word "quality" took on a more encompassing meaning, which included important software factors such as reliability, maintainability, portability, and robustness. In addition, the pursuit of software quality, when applied to interactive computers, required that a heavy emphasis be placed on the human aspects through design of "user friendly" computer features. Thus, software quality came to embrace many of the product integrity disciplines under a single banner.

1.4   Back to Square One

During the historical development of the assurance sciences, it became abundantly clear that the human element is extremely important in assuring product integrity. In the early days, the master craftsman's pride was key to achieving this objective. The Japanese have successfully resurfaced this ideal and have accomplished significant gains in all areas of product assurance, such as quality and reliability. In the United States, we are rekindling the pride factor and restrengthening our commitment to product integrity. Quality Circles are springing up to provide a group focus on the solution of product integrity problems and issues. An even more innovative approach perhaps is the use of a sociotechnical systems (also called high-performance work systems) organizational concept in which an organization is designed into self-managed work groups. Within these teams, employees work toward a common purpose and are given control over day-to-day decisions. The belief is that with more control over the decisions that influence their working lives, employees will produce superior results in the form of greater productivity and quality. Whatever the approach, the thought is the same—to get all workers to have pride in their work and to commit themselves to product integrity as in the days of the master craftsman.

1.5   Arts to Sciences

From the above, it is plain to see what has happened in the historical growth of product assurance. Disciplines grew to fill the voids not filled by other disciplines. Quality control took care of the quality characteristics of the product. But how about the product's durability? Well, reliability came along to fill that void. Still, when a product failed, repairability had to be considered. Maintainability and human factors filled that void. But was it safe to use? Maybe not—hence product safety. And so it went. Today we have even more related disciplines that are concerned with assuring product integrity: integrated logistics support, systems effectiveness, etc. The tree of product assurance (see Figure 1.1) has grown to the point where it is strong and viable, its many branches nurtured by strong roots.

Another obvious factor in the growth of the assurance sciences, but certainly worthy of repetition, is the role of the military or military products. In the early days, the need for weaponry spearheaded the fundamental beginnings of the assurance sciences. Later throughout its growth, the assurance sciences were continually pushed ahead through the concern of the military. Specifications, standards, and engineering/statistical techniques were developed specifically by the mili-

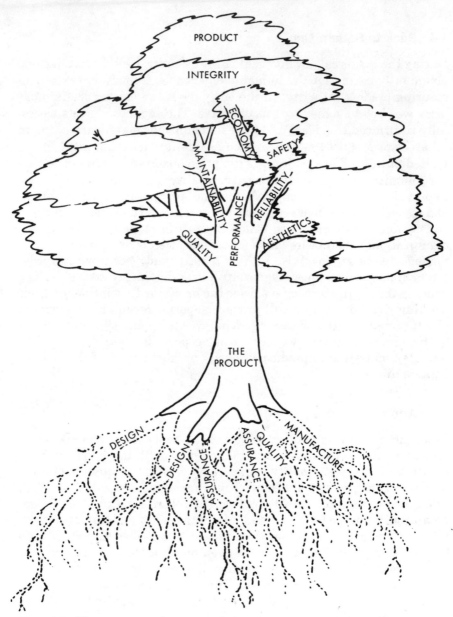

Figure 1.1 The assurance sciences.

tary to address the problem in one facet or another of product integrity. Commercial products contributed to this growth, but it was primarily from the viewpoint of "riding the coattails" of the military. The need was more critical. In many cases it was truly a battle—a battle of survival!

Over time, we have seen an orderly transition of the assurance sciences from arts to sciences. In the beginning, we had craftsmen who employed highly developed skills to make their products. These craftsmen possessed the necessary aptitude, dexterity, and ingenuity to make the products of their day, but as products became more complex and intricate, and were produced in greater volume, these characteristics were insufficient to meet the challenge of assuring product integrity. Furthermore, the product assurance approaches were inadequate for the increasingly difficult task at hand.

Today we have progressed to a point that assuring product integrity is indeed a science. Science has been defined as "a branch of knowledge dealing with a body of facts systematically arranged and showing the operation of general laws." The disciplines of product assurance (such as quality control, reliability, maintainability, product safety) employ scientific management and technical methods, techniques, principles, concepts, and approaches as they go about their business. In the remainder of this book, we present each of these product-assurance disciplines and show how they are integrated to assure product integrity.

2

Elements

All are but parts of one stupendous whole.
ALEXANDER POPE

2.1 Product Integrity

There does exist a set of engineering functions that are necessary to the assurance of a "good" product—i.e., product integrity—that is independent of who performs those functions or how they are performed.

In order to understand these engineering functions, we must know what product integrity is. Product integrity consists of a predetermined optimum balance of (see Figure 2.1) performance; aesthetic appeal; reliability; ease, economy, and safety of maintenance; and consistency— all at a given cost, of course. From the user's point of view, these product attributes can be simply defined as:

Functional performance—the ability of the product to do what the user wants it to do

Aesthetic appeal—the ability of the product to please the human senses

Reliability—durability, longevity, and infrequent malfunction

Ease, economy, and safety of operation—self-explanatory

Ease, economy, and safety of maintenance—self-explanatory

Software interaction—characterized by reliability, correctness, and efficiency, while being flexible and maintainable

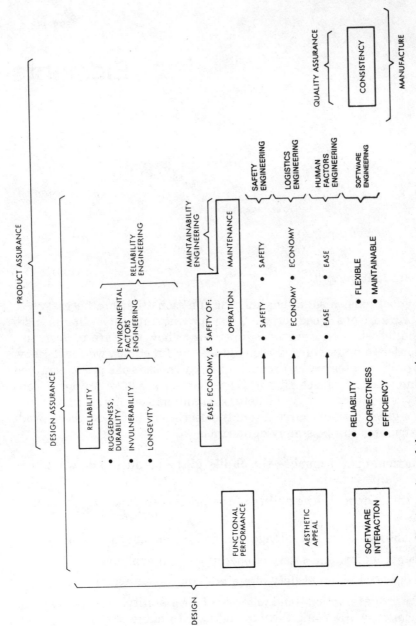

Figure 2.1 Product integrity and the assurance sciences.

Consistency—that quality of the product ensuring that any one copy of the product (and its constituent parts) is dependably the same as any other copy

2.2 Design Assurance

Notice that, given consistency (which is primarily a function of manufacturing), the other six elements of product integrity are functions of the product design. Given that the primary function of design creation is performed, design integrity is assured by *design assurance,* which assures that design creation proceeds in a systematic, orderly fashion achieving optimum design solutions for performance, aesthetic appeal, and the four remaining elements of design integrity.

2.3 Performance/Aesthetic Appeal

Both performance and aesthetic appeal are pure design creation attributes. It is a given that *all* design attributes are the basic responsibility of design engineering in the design creation process. While performance is measured by design parameters covered by the assurance disciplines, aesthetic appeal remains almost completely outside these disciplines. Design assurance evaluates, ensures, and documents the degree to which the designed product meets defined performance requirements, particularly as described in the remainder of this chapter.

2.3.1 Reliability

Reliability is assured by reliability engineering:

1. *Systems reliability engineering.*   Assures that realistic, measurable reliability requirements are established and that the product, as designed, will meet these established requirements.
2. *Components reliability engineering.*   Assures that the most appropriate parts and materials are selected for use in the design and that these parts and materials are properly applied in the design.
3. *Environmental factors engineering.*   Assures that the product, as designed, successfully "lives with" its anticipated use environment.

2.3.2 Ease, economy, and safety of operation

These considerations of ease, economy, and safe operation are a function of:

1. *Human factors engineering.*   Assures the provision of an operator-compatible product design.
2. *Logistics engineering.*   Assures minimization of requirements for operating consumables, and the availability and economy of necessary consumables.
3. *Safety engineering.*   Assures that the design minimizes hazards to the product, the operator, and the casual passer-by.

2.3.3   Ease, economy, and safety of maintenance

Maintainability engineering assures that the product design embraces features that allow for maintenance to be accomplished in the least possible time with a minimum of resources—given that *human factors engineering* assures the existence of a compatible repairer-machine interface in the design; *logistics engineering* assures provision for adequate and economical field support; and *safety engineering* assures minimization of any safety hazards that might be found across the repairer-machine interface.

2.4   Software Interaction

This important aspect is addressed by the software quality assurance function through the following:

1. *Software quality management.*   Assures that the software requirements reflect the user's quality expectation and provides a continuing, independent review and audit of the software development process.
2. *Software verification and validation.*   Assures that the developed software meets the specification requirements, as well as the user's perceived expectations of quality.
3. *Software maintenance.*   Assures that software changes actually correct problems or enhance performance without introducing (other) software problems.

If all of the above assurance functions are properly executed, the results will yield a total product design, i.e., a design for a product that will possess integrity.

2.5 Manufacturing Quality Assurance

Before the actual product exists, it must be manufactured. It is the mission of the manufacturing function to fabricate the product as designed. In other words, to achieve consistency, which is assured by *quality assurance*—seeing to it that the as-designed integrity of the product is not degraded by the manufacturing process. Quality assurance comprises two "control" functions.

1. *Quality control engineering.* Establishes process precision, workmanship, and inspection skill criteria, nature and number of product inspections; accept/reject criteria; etc.—and provides the motive force for corrective action in the manufacturing process, as the need is indicated by quality control inspection.

2. *Quality control inspection.* Executes the established inspection procedures.

So, given that the primary functions of design creation and product fabrication are performed, the integrity of the resulting product is assured by the *assurance sciences*—design assurance and quality assurance, whose elements are delineated above.

It should be noted that the term "quality assurance" is used here in the context of both a design assurance element (namely software quality assurance), as well as a manufacturing assurance related element. When used by itself, quality assurance will relate to the manufacturing assurance activity. However, where there is the possibility of confusion, the descriptors "software" or "manufacturing" will be added as required.

Now that we understand each of these elements in terms of succinctly stated individual objectives, let us proceed toward a more comprehensive understanding of these disciplines, as they now exist.

2.6 Design Assurance Elements

Design assurance is the process (or management mechanism) through which the creation of a satisfactory and complete design takes place. Design assurance is synonymous with design review in its larger sense and includes design review in its narrower sense (i.e., formal examinations of the design). It seeks to make sure that nothing is forgotten or ignored, and that the total design includes not only the engineering drawings and specifications necessary to physically and functionally describe the intended product, but *all* documentation required to support the manufacture, test, delivery, use, and maintenance of the product. Design assurance embraces all of the design-related assurance sciences.

Reliability engineering influences the design to describe a durable, long-lived product which rarely malfunctions. It seeks to minimize the frequency of component failures and to mitigate the effects of component failure on product function, *by design*. To this end, reliability engineering attends to the appropriate selection and proper specification of component parts, assurance that they are conservatively stressed under all anticipated use conditions, and assurance that the assemblage of functional components represents the optimum design solution for reliable product performance. In simpler terms, this means assurance that satisfactory designed-in product reliability is achieved, given that individual component parts are properly selected and applied. This is the mission of what we have called system reliability engineering. This function is traditionally performed by mathematically synthesizing the product's proper operation (or malfunction) as a function of the statistical laws of chance and historical experience. In this manner, various design alternatives can be compared, weak links discovered, and an optimum design for reliability achieved. The measure of success can be a statistically designed test of a representative sample of the product.

Components reliability engineering (which seeks to assure the validity of the assumptions necessary to the system reliability engineering function) sometimes performs the same function but in microcosm. That is, the system (product) becomes a component part; component parts become the physical and chemical mechanisms that lead to component failure; and the laws of chance are functions of stress. More commonly, however, historical component reliability data are developed by long-term testing. Conservative component parts application is encouraged by the establishment of derating rules and monitored by means of stress analysis.

Reliability engineering requires the same skills as those required for product creation (electrical, aeronautical, mechanical, chemical engineering, etc.), with emphasis on analytical skills and statistics, plus systems analysis skills and experience with the component parts of interest.

Environmental factors engineering is concerned with assuring that the product, as designed, will withstand the rigors of any anticipated environmental stresses, within any required safety margins. This engineering function includes analysis of structural stress/strain, thermal characteristics, electromagnetic interference radiation/susceptibility, nuclear radiation vulnerability, etc. In addition, an important aspect of this function is evaluation testing. Evaluation testing requires skills in environmental and sometimes rather exotic functional testing techniques. The goal is to verify using product samples, that all of the product to be produced will withstand the use environment. Tests involving measurement of functional operation under environments of sun, rain, wind, immersion, altitude, temperature extremes, shock, vibration, etc., in any combination, are typically accomplished. Evaluation tests are invariably assurance tests directed at the product in general, and may or may not be verification tests. Figure 2.2 shows these relationships. The main objective in evaluation testing is to duplicate use conditions as closely as possible for assurance of product class conformance. Evaluation testing interfaces with practically all the engineering functions, including reliability, maintainability, and safety engineering.

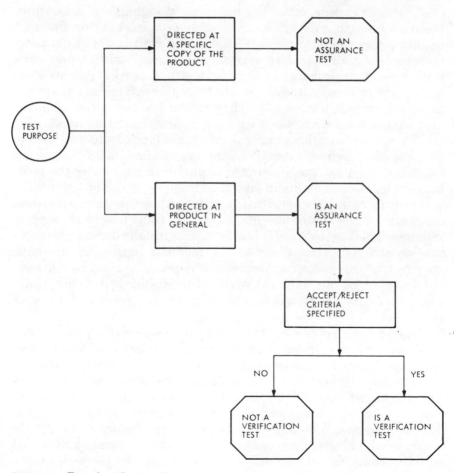

Figure 2.2   Test classification flow chart.

Human factors engineering is aimed at making sure that the product is designed to fit the requirements and abilities of the user. Its major implement is product design analysis from a people viewpoint, e.g., reach required, weight to be lifted, etc. An important facet of human factors engineering is human *habit*. For instance, people *expect* to achieve an increasing function by turning a knob clockwise, except in the case of fluid control valves—then they expect the opposite. Human factors analyses account for these peculiarities and guide the design accordingly. The analyses cover the people aspects of use and maintenance. A variety of handbooks, manuals, and the like exist to define the "standard individual," i.e., motives and sensory abilities, likes and dislikes, and their statistical distributions within certain population classes. The human factors engineering discipline is known by at least one other name—personnel subsystems.

In the past decade, there has been an awakening realization that the most reliable, safe, and maintainable product, built by the most quality conscious company, is of little use without the proper support documentation, consumables, service, and spares. Logistics, provisioning, spares engineering, and integrated logistics support (ILS) are some of the more common names given to a discipline established to address these assurance issues. As a discipline, it is rather new, although pieces of it have existed for decades, e.g., spare parts, operating handbooks, etc. The logistic function's raison d' etre is the lack of attention paid to the cost of a product's *use*. The function is aimed at providing an organized, planned, and integrated approach to supporting the product during use—at minimum cost. Most frequently, ILS has evolved in a *management* integration function, bringing together the appropriate elements of the assurance disciplines to satisfy product support requirements. The results of ILS activities are usually documented product support data, i.e., maintenance manuals, operating manuals, sparing analyses and recommendations, depot data, etc. The considerable amount of the technical work accomplished in reliability and maintainability engineering is utilized by an integrated logistics program.

The importance of safety engineering, at least from the point of view of protecting the operator from the machine, is obvious. Not so obvious, perhaps, is the safety engineering concern with protecting the machine from the operator, and from itself. The aim of safety engineering is to assure that the product is designed so that it offers minimal risk to the physical well-being of the user, the surroundings, and itself. The basic approach of safety engineering is to identify and critically examine all high potential energy sources within the system composed of the product and its surroundings, including the people in those surroundings. This analysis includes all product states—operation, failure, repair, storage, handling, and transportation. Even the simplest examples of high potential energy—e.g., the energy potential at a sharp corner with a 200-lb human body falling against it—must be accounted for. In practice, safety engineering is usually performed by examining the product design in terms of checklist criteria which have been formulated from prior analyses of likely high-energy sources and from sad experience. Safety engineering requires considerable common sense plus the engineering skills required for creation of the product. Although design analysis is relied upon as the assurance of product safety, verification of safety achievement is also typically secured by testing. For example, color television sets are tested to ensure that radiation emittance is within safe levels established by the government.

Although a perfectly reliable product would never fail or wear out, such is not the real world. Products *do* fail and *do* wear out. For most products, failure does not mean the end of that product's life—that is, repair is possible. Maintainability engineering is concerned with the ease and economy with which a product can be restored to (corrective maintenance) and/or kept (preventive maintenance) in an operable condition. Wearout constitutes the end of product/component life. Maintainability engineering seeks to assure physical partitioning of the product in such a manner that product components subject to consumption or wearout are economically replaceable and that the longest possible lives for such components can be achieved through preventive maintenance. Assurance must exist that the product is designed to be maintained in some reasonable maintenance environment which exists, or will be provided, for the product. This maintenance environment, anticipated by the product design, includes maintenance personnel qualities and skills, available test equipment and tools, maintenance manuals, special training requirements, spare parts, supply lines, and administrative difficulties. Therefore, given a fixed maintenance environment, the product must be designed to accommodate this environment. In cases where the producer has some control of the maintenance environment, maintainability engineering must sort out the cases where the design must accommodate the environment, and those cases where provision must be made for the environment to accommodate the design. Ideally, maintainability engineering requires skills in logical and statistical analysis, design engineering skills, and experience in actual maintenance. An understanding of the principles of human factors and reliability engineering is also necessary since maintenance personnel are human and maintenance frequency is a function of reliability. Maintainability is a measurable design parameter. It can be specified, predicted, and measured quantitatively and/or qualitatively— in real, simulated, or imagined maintenance environments.

2.7 Software Assurance

The past two decades have produced, as one of the developments of the information age, an amazing array of electronic products. The plethora of commonly used electronic systems and devices is increasingly dependent on two new "parts"—software and firmware. Software programs, either "burned" into an integrated circuit (firmware) or accessed as code by a system (software), are becoming as much a part of delivered products as the "hard" goods themselves. In fact, in the last decade the cost of software engineering (in dollars) of electronic systems has risen to more than 50 percent of the total. The implications of this figure cause real concern for the assurance disciplines. Every design assurance and quality assurance issue exists in the development of software that exists in the design and manufacture of hardware.

The significant difference is that no set of standard, accepted techniques is commonly available which will give equivalent assurance that a product containing software completely meets defined reliability, maintainability, human factors, safety, etc., requirements. For that matter, even requirement definition in the software arena has not completely matured.

An understanding of the issue can be gained by viewing the production of usable software as analogous to the production of hardware. Software must be designed to requirements in much the same way as hardware is. In the case of hardware, design data (drawings, specifications, direct computer-aided design data—CAD, etc.) determine what is to be manufactured. In the case of software, after design, there is a coding step—roughly equivalent to the hardware drafting and design documentation activity. At this point the two activities differ. For hardware, a new organization—manufacturing—starts producing the hardware. For software, the activity (at least currently) generally remains the responsibility of design engineering. Usually, it is not until the software and hardware are integrated, that the software joins the manufacturing process.

At present, industry is trying a variety of approaches for assuring the total integrity of products that incorporate software. The closest activity to an assurance science exists in an area called "software quality assurance," which in this book is treated as a design assurance element and is covered separately in Chapter 13.

2.8 Manufacturing Quality Assurance Elements

The foregoing assurance sciences, when performed in concert with each other and the other elements of the design process, give assurance that a *total design* exists—a design that embraces all parameters of "goodness" of the product, within the full product life cycle environment. It remains, then, to convert this paper design to the physical/functional product it describes. That this conversion takes place properly and consistently is seen to by a function (management mechanism) we shall call "quality assurance." Quality assurance is to product manufacture as design assurance is to product design. Its mission is to assure that the design characteristics of the product are not degraded by the manufacturing process—in other words, to assure that each model, or copy, of the product conforms in every way to the intent of the design. Quality assurance is synonymous with "quality control" in its larger sense and includes quality control in its narrower sense—systematic monitoring and control of the materials, processes, and workmanship of manufacture. Quality assurance comprises quality control engineering and quality control inspection functions.

Quality control engineering serves to devise practical working procedures for achieving specific goals established by the quality assurance function. It establishes manufacturing operator and inspector skill and training requirements, the nature and points of inspections, sampling and accept/reject criteria, etc., and provides the direction for correction of deficiencies in the manufacturing process. The skills required for quality control engineering vary considerably with product type, but generally consist of systems analysis, statistical, and metrological skills, coupled with experience in the manufacturing processes of interest and in quality control inspection.

Inspection, as we speak of it here, should not be thought of as merely visual. It includes all operations necessary to detect and measure conformance of the product to predetermined standards, e.g., electrical test, x-ray, flatness gauging, and airflow measurement. The inspection and data reporting procedures established by quality control engineering are executed by the quality control inspection function. Inspector skill requirements, as we have said, are established by quality control engineering (task by task) and consist generally of visual/tactile acuity, ability to use test instruments, and attention to detail—coupled with experience as manufacturing operators or laboratory technicians and specialized training as inspectors.

2.9 Product Assurance

The preceding generally recognized, but somewhat amorphous, body of related engineering disciplines (assurance sciences) form the entity we have called product assurance. Actually, the assurance sciences are necessary, but not sufficient, to form a practical product assurance function. The effective assurance of product integrity requires, not only the

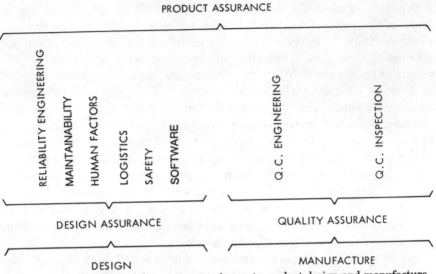

Figure 2.3 Relationship of the assurance sciences to product design and manufacture.

engineering techniques represented by the assurance sciences, but some coherent direction or management of their application to the product design and manufacture. It should be clear from the previous discussion that the technical elements can be logically related, in some way, to the process of creating a product. Figure 2.3 illustrates the idea that these elements are neatly separable into (1) those that are related to

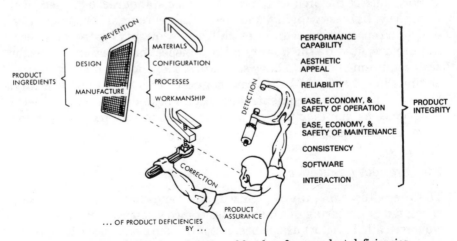

Figure 2.4 Product integrity as a function of freedom from product deficiencies.

product design and (2) those that are related to product manufacture. Figure 2.4 recognizes the commonality of purpose and approach of the assurance sciences across both design and manufacture, i.e., *preventing, detecting,* and *correcting* product deficiencies to achieve product integrity. Now, if we can somehow incorporate both of these somewhat dichotomous ideas into a single structure, we should have a logical beginning for an effective management structure.

3

Unification

In necessarius, unitas
In things essential, unity

RUPERTUS MELDENIUS

3.1 Current Status

The assurance sciences provide the disciplines required to assure product integrity. We have defined this application as product assurance. But, to the world at large, what *is* product assurance? There are almost as many answers as there are people in the industrial world. A frequent answer from nonspecialists is, "quality control." The usual answer by an ambitious manager of an assurance function is: "marketing, design engineering, quality control, drafting, plant maintenance...." The definition of product assurance varies from company to company and person to person. Further, the definition is invariably in terms of some specific and noncomprehensive set of assurance sciences—rather than in terms of the goal to achieve product integrity.

Separatism of the assurance sciences has fostered narrow specialization—resulting in loss of motivation, engineering perspective, peer respect, and self-respect. One serious effect of this has been to make the truly qualified product assurance manager a very scarce animal. Who will bring the unified product assurance function into being? The individual who has spent a lifetime in an autonomous quality assurance organization? Not very likely! Even the most well-rounded chief engineer is unlikely to have the necessary qualifications for effectively directing the product assurance function for a complex business or product.

The goals of product assurance are user-oriented. Also, product assurance functions are basically "see to it" rather than "do it" functions. These two ideas capture the essence of the product assurance unification problem. Product assurance is paradoxical in that it is the consumer's advocate, while it lives with and is responsible to the producer. It is an extension of management ("see to it"), yet requires a rather extensive body of technical expertise (a considerable amount of "do it" is necessary to the product assurance "see to it" functions).

It is often difficult for the producer's top management to see enough self-interest in product assurance to warrant any significant investment of resources. A splinter of product assurance is, therefore, very often content just to exist, wherever it finds itself in an organization. Then too, practitioners of the assurance sciences all too often lose sight of the legitimate profit motive of the producer. Their attempts to sell their various "cults" usually amount to taking the position that the tax must be paid lest the customer community put us all in jail or that one must keep up appearances—like going to church every Sunday. Perhaps the most serious obstacle to effective application of product assurance, however, is the lack of any unified, coherent, and consistent approach. Proponents of the individual assurance disciplines speak in a Babel of tongues and are forever at each other's throats over pieces of the corporate carcass already ripped off by one of their number.

In spite of the crazy-quilt history, divisiveness, and incoherence of the assurance science evolution, most aspects of product integrity are being pursued—with varying effectiveness—in most segments of industry. The traditional assurance functions (sometimes recognized as such, and sometimes not) are formally present, or absent, in various combinations—from company to company or project to project.

Quality assurance, or quality control, is a formal function in virtually all manufacturing companies. It is usually applied to the total product—software, documentation, and hardware—incoming as well as outgoing. Quality assurance goals are well understood and in consonance with the views espoused by this book. Quite frequently, some aspects of reliability engineering and environmental factors engineering are performed under the mantle of quality assurance.

Design assurance, also, is almost universally formalized, at least to the extent of a design review board or committee or some mechanism to assure the orderly achievement of an acceptable solution to the design problem. A comprehensive understanding of what constitutes the design problem is somewhat hit-or-miss. Functional performance and aesthetic appeal are always considered. Consideration of reliability and ease, economy, safety of operation, and maintenance is spotty. These considerations are sometimes left to the design engineer (with or without review) and sometimes assured by formally constituted functions such as reliability engineering, system/product safety engineering, etc., which were enumerated and defined in the previous chapter. All of these design assurance functions are occasionally formalized within a single company, but seldom, if ever, into a single organization.

3.2 Reasons for Unification

The most obvious reason for unification of the assurance sciences into a coherent functional unit is their singleness of purpose and similarity of technique. The singleness of purpose is made obvious by the preceding chapter. A review of the descriptions of the product assurance elements in Chapter 2 will verify the similarity of techniques.

As matters now stand, the relative inefficiency of organizationally scattered groups of assurance specialists is a problem. We see cases where reliability engineering goes to great trouble and expense to achieve accurate estimates of equipment reliability, but *not* for the specific hardware subdivisions of interest to the maintainability engineer or to the spares planning people. Maintainability engineering, systems safety engineering, and human factors engineering may come up with competing solutions to a maintenance safety hazard, *or* they may all ignore it, assuming it to be someone else's responsibility. Quality assurance may "solve" a "quality" problem by having the factory select (parametrically) a troublesome piece-part, although the proper action may be to solve a design problem.

Foreign competition is an ever-growing reason that the assurance sciences *must* be unified into an effective, efficient promoter of product integrity. "Made in Japan" is no longer synonomous with "cheap junk." Japanese products have, in a relatively short time, achieved an enviable reputation for inexpensive quality (meaning product integrity). That they have achieved this is due, in large measure, to an astute unification of the assurance sciences—including the assimilation of appropriate functions into the design and manufacturing processes. Similar phenomena are occurring around the globe—to the extent that "made in U.S.A." is in danger of becoming synonomous with "expensive junk." Admittedly, there are other factors at work here than the different methods of implementing the assurance sciences. Product assurance is a factor, however, and we predict it will become more important as time goes on.

In any case, we have reasons enough here in the United States to be shy about producing or using shoddy and expensive products. Of late, producers of consumer goods have been especially vulnerable. Product liability claims have increased tremendously in the last few years— showing no signs of decreasing. Awards have grown in number, amount, and variety. Of particular significance is the change in interpretation of the law. Personal injury claims having any reasonable basis almost invariably are winners. One case, *Henningsen* vs. *Bloomfield Motors,*[1] established the landmark precedent of not requiring proof of negligence to win a personal injury product liability award. Ten days after Mr. Henningsen bought a new Plymouth, when his wife was driving the car, it allegedly went out of control and struck a stone wall—injuring his wife, damaging the car, and, presumably, the wall. Mr. Henningsen sued both Chrysler and Bloomfield Motors. A negligence count was not sustained by the evidence—but the trial court instructed the jury on breach of implied warranty of merchantability, and the jury held both defendants liable. This case clearly shows judicial attempt to protect consumers and established that the disclaimer in the automobile express warranty is not all-inclusive because of the unequal bargaining position of a consumer and the manufacturer or dealer. Strict liability is now the rule, not the exception. Claims are now being won for faulty products which have no loss other than money. For example, *Santor* vs. *A. and M. Karagheusian, Inc.,*[2] which involved a Gulistan rug purchased from a dealer and not the manufacturer, and which soon revealed flaws, was directed against the manufacturer. The New Jersey Court found for the plaintiff's economic loss, i.e., the difference between the present value of the carpet and its worth had it been a good carpet. Thus, as far as the individual consumer is concerned, adequate legal protection now exists and gets more inclusive all the time. Producing and selling a product having less than adequate integrity is a serious thing. It does not follow that one may therefore be cavalier about the assurance aspects of acquiring and using products. This is particularly true of industrial users who purchase parts and materials to put their product together. They are normally expected to specify exactly what they want and to take the steps necessary to assure that they get it. The same rules of liability suits do not appear to apply in their case. Whereas the landmark case of *Henningsen* above was won by the individual consumer partly on the basis of "inequality of bargaining," attempts to use this theory by industrial users have not been successful—it is evidently assumed that the industrial purchaser (user) is a thoroughly sophisticated one. It appears that contractual clauses dealing with defects in workmanship, inherent design flaws, etc., supply the only substantive basis for redress in the industrial arena.

Federal government contracts present yet one more area for concern with user protection and redress. In this case, it is the government that is the user. Considerable leverage is available to the government, e.g., implied threat of no additional contracts, imposition of more extensive inspection and control, ability to review costs and activities in the producer's plant to practically any depth, or contract renegotiation. The government as a user has not yet successfully employed product liability suits. One viewpoint is that contractors working under a government contract should not be held liable for their products because of the limitation of profit and fee and because of the government and/or military standards imposed on them. The most recent development in government (user) protection is the 1984 federal law requiring warranty clauses in such contracts.

In addition to the possibility of direct economic loss due to accepting inferior components, there is always the danger of feed-through to the end product—bringing us back to the specter of being on the receiving end of product liability action.

3.3 Unification Approaches

The many compelling reasons for unification of the assurance sciences began to be generally recognized in the mid-1960s. Some progress has been made. Proper manufacture is now generally assured by organizations which have been upgraded (in function, as well as in name) from quality control to quality assurance. The concept of total quality control has been preached, if not widely practiced. In the design side of the house, we have seen various combinations of the design assurance functions being unified and given such titles as design assurance, systems assurance, or systems effectiveness engineering. Certain projects have carried a product assurance project manager (with and without supporting staff) with the "see to it" responsibility and authority for total product integrity on that project. It has been our observation that this is an effective but not necessarily efficient approach.

All of the organizational attempts to increase the effectiveness and efficiency of the assurance sciences can be classified under one of two basic approaches (or some combination of the two):

1. *The function-oriented approach.*   One attempts to collect and develop "skill centers" of the various branches of the assurance sciences. These individual skill centers may or may not be organizationally tied together under a product assurance mantle.

2. *The project-oriented approach.*   More broadly based assurance science groups (or individuals) become more or less homogeneous with the design-production team for individual projects or product lines.

Like most other ideologies, pure forms of either approach seldom occur in nature, although we have seen a few instances of almost undiluted function-oriented organizations. Most organizations try to strike a compromise which avoids whatever disadvantages they consider intolerable. The salient advantages and disadvantages of the two basic approaches are tabulated in Table 3.1. This tabulation presumes a rather specific industrial environment. It assumes the expenditure of significant product assurance resources to assure the integrity of a reasonably diverse or complex product line. Thoughtful consideration of relative effectiveness in achieving the various goals of the table leads to several conclusions:

TABLE 3.1   Product Assurance Effectiveness versus Organization Approach

	Product Assurance Orientation	
A Sampling of Product Assurance Goals	*Function Oriented*	*Project Oriented*
Establishment of consistent P.A. policy and procedures	Good	Poor
Definition of required integrity for each product	Good	Better
Design deficiency prevention	Good	Better
Early detection of deficiencies	Good	Better
Isolation of deficiency root cause	Better	Good
Development of corrective action	Good	Better
Implementation of corrective action	Good	Better
Prevention of recurrence of like deficiencies	Good	Poor
Documentation of evidence of product integrity	Better	Good

1. For modest expenditure of resources devoted to product assurance, the project-ized approach seems to be more efficient than the functional approach.

2. For more ambitious product integrity goals and more dedication of resources, the functional approach appears to become more efficient.

3. There should exist some optimum approach which somehow takes advantage of the effectiveness of both approaches and suffers the least possible inefficiency.

This kind of thinking allows one to construct a family of curves such as Figure 3.1. This family of curves displays the conclusions reached earlier. In addition, it is shown that *some* devotion of resources to product assurance would result in design and production cost savings *and* improved product integrity. It also follows that for the single-product-

line company either approach secures the advantages of both, i.e., there
is no difference between the two approaches.

Note that either of the two basic approaches, as well as their combinations, may incorporate various degrees of organizational unification.
Very recently, we have begun to see product assurance responsibility
and authority vested in a single entity. Nevertheless, even those product assurance organizations almost universally suffer several serious
flaws—lack of overall design review responsibility and authority; lack
of overall product integrity evaluation capability; inadequate contact
with the user; and noncomprehensiveness because of failure to include
all necessary assurance sciences and because of dilution with nonassurance functions.

3.4   Effective Unification

Having reviewed the product assurance goals, the essential reasons for unification of the assurance sciences, and some of the recent approaches toward

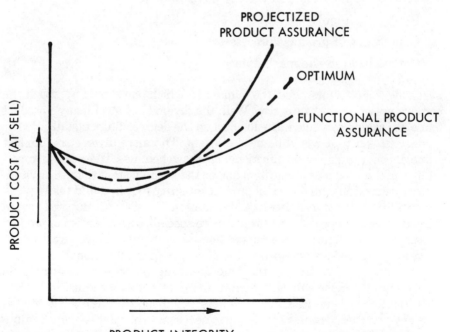

Figure 3.1   Product cost vs. product integrity.

more effective product assurance—where are we? Well, it appears reasonable that the assurance sciences can operate best if they have a unified direction. Certainly there is sufficient and growing need for better product assurance. Thoughtful, knowledgeable, and energetic people have made considerable progress toward this end in the recent past. And yet, there is no universal recognition of the "one-ness" of the assurance sciences.

We must admit that there appears to be no single organizational arrangement of the assurance sciences that will serve everyone, for all time. Nevertheless, there is a single, rational approach to assuring product integrity. As for the achievement of *any* set of goals, one must be concerned with motivation, organization and planning, and the necessary resources. We considered the subject of motivation earlier in this chapter, at least to the extent of satisfying ourselves that some unification is of value. Organization and planning are the subjects of the chapters immediately following. Nevertheless, some general organization and planning guidelines are in order here.

We have inferred that effective unification of the assurance sciences depends on a number of factors:

Diversity of product line

Product application and the use environment

Product volume

Complexity of products/processes

Competition in the marketplace

How do these factors affect the manner in which one should approach the assurance of product integrity? Well, the diversity of a company's product line, for example, would have an effect on the degree of functionalization or projectization that would be most efficient. The more diverse the product line, the more efficient the functional approach becomes. The relative importance of the product's application and/or the peculiarity of the use environment determines the degree of product integrity required and the requirements for integrity verification/documentation. Thus, we are directed toward the functional approach and toward more specialized application of product assurance techniques. Price competition tends to pull us away from proliferation of specialists and toward the efficiency of the functional approach.

Industry recognition of the true meaning of product assurance is necessary to more efficient achievement of better product integrity. Unification of the assurance sciences promises to make greater efficiency possible. Greater efficiency is necessary in order to meet competition, both foreign and domestic, and to avoid product liability actions. Product assurance unification consists of organizing the proper assurance sciences by function, by project, or in some optimum combination. The structure and practices of the ideal product assurance organization depend on the precise nature of one's business.

3.5   Notes

1. *Henningsen v. Bloomfield Motors,* 32 N.J. 358, 161 A. 2d 69 (1960). 2.
2. *Santor v. A. and M. Karagheusian, Inc.,* 44 N.J. 52, 207 A. 2d 305 (1965).

4

Organization

*The organization to some degree remakes
the individual, and the individual to
some degree remakes the organization*
                    E. WIGHT BAKKE

4.1 Background

The previous chapter details the reasons and general approaches toward
unification of the assurance sciences. The next logical step is to develop
a specific structural relationship between the assurance functions, i.e.,
to develop an organization. A useful question to examine is why there
needs to be any organization at all. This small anecdote might help in
understanding. Probably the first organization was conceived when Sam
Club arrived at his cave after the third or fourth skirmish with a saber-
tooth tiger. His clothes were in worse-than-usual tatters, and he was
fed up with being knocked around by the animals. "After all," he said
to himself, "I'm a human! I'm supposed to be able to outsmart any old
animal around." After some thought, the light suddenly dawned. "Why
not get Jake Spear to go with me the next time I go out?" he thought.
Then his thought expanded further, "Why not get Matt, Hugo, Killer,
and the rest? I'll *organize* them into a hunting party!" The results were
predictable. They got their tigers with not nearly as much bodily abuse.
A group with a common goal will achieve more than the same number
of people working individually toward the same goal.

The assurance sciences do have common goals, and an organized
approach toward their achievement will be more effective than individual actions. This organized approach does not occur simply as the

result of structured functions. There must be people in the functions motivated to achieve the functional goals. Even that is not sufficient. Each person (or organizational subdivision) must have specific responsibilities, and informed leadership must be provided. Authority and responsibility for the various aspects of product integrity must be clearly and unambiguously defined for each organizational unit, as well as each person within the structured organization. However, some structure is essential. These structures can be as simple as Sam Club's (in small businesses) and very complex—as might be found in a large corporation.

4.2 Corporate Role

The type of assurance organization in a given company will be determined almost exclusively by that company's top management personnel. How much "do it" function the organization will have, the balance with a "see to it" function, and the amount of integration with other operating organizations will be decided by just a few top managers.

The effectiveness—maybe the very existence of an assurance organization—will be a direct result of corporate recognition of the need for assurance disciplines. This may be obvious, but it is necessary to understand. If there are no top management personnel with assurance backgrounds, strong interest, or experience, a weak assurance organization generally results. This situation is more pronounced if there has been no previous history of an effective assurance organization within the company. The work experiences, orientation, and biases of the top management operating staff play an overwhelming role in assurance organizational policy and structure.

Another factor, which significantly affects corporate attitude, is the company's experience with their customers. If the customer is a household consumer, some quality pressure will be exerted. If the customer is a government agency, an overwhelming plethora of governmental specifications will exert considerable pressure toward having an effective and recognizable assurance organization. The following list of governmental specifications in assurance areas gives some idea of just how much requirement documentation can exist (all of these have specific assurance organizational requirements):

MIL-S-52779   "Software Quality Assurance Program Requirements"

MIL-STD-785   "Reliability Program for Systems and Equipment Development and Production"

MIL-STD-470   "Maintainability Program Requirements (for Systems and Equipments)"

MIL-STD-1472	"Human Engineering Design Criteria for Military Systems, Equipment, and Facilities"
MIL-Q-9858	"Quality Control Requirements"
MIL-STD-882	"System Safety Program for Systems and Associated Sub-Systems and Equipment; Requirements for"
MIL-STD-1369	"Integrated Logistic Support Program Requirements"

Government regulatory agencies, such as the Food and Drug Administration (FDA), exert direct pressure on company management to have an effective assurance organization. A past court decision upheld the FDA's requirement that a delicacy food canner have full-time quality control personnel before allowing him to distribute his product. Surveys are made by military services prior to contract award to ensure that the potential contractor has an acceptable quality (and sometimes reliability) organization.

The consumer advocates have also put major pressure on company managements to recognize the quality-safety-reliability-maintainability needs of the public. This has translated into considerable assurance organizational development within businesses producing consumer products. Consumer hotlines, surveys of customer's gripes, and like activities have now become part of the corporate assurance function—although not necessarily part of assurance organizations.

The rising flood of product and service liability lawsuits has put economic pressure on company management. A large number of these lawsuits are tried with both the defense and the prosecution basing their case on elements of the assurance functions. Many have been won or lost strictly on the basis of whether an effective assurance organization exists—with the normal assurance documented records. Corporate managements do heed these economic pressures and are recognizing the increasing need for effective assurance activities.

Even where the federal government is almost a complete working partner with industry in equipment development and production (defense contracting), a recently enacted law requires specific warranties. Again, this provides emphasis on assurance functions and organizational considerations.

Finally, the very heavy emphasis on assurance disciplines and their organization by highly competitive non-U.S. companies (e.g., Japanese) has heightened U.S. corporate management's interest and concern in assurance organizations.

All these factors make organizational structure of assurance functions of significant interest.

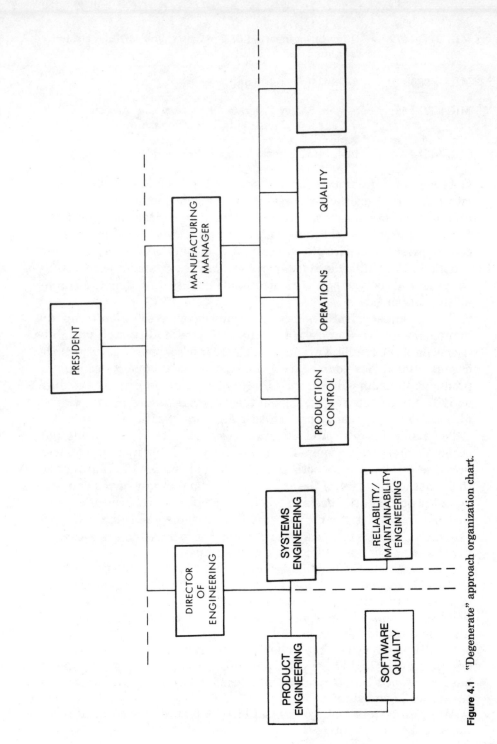

Figure 4.1 "Degenerate" approach organization chart.

4.3 Structural Possibilities

Organizational structures exist in almost infinite variety. Perhaps the least formal structure could be represented visually by a scatter diagram of names with no hierarchal structure, i.e., no relationship among functions. The most formal might be represented by the typical "family tree" with each block representing a strict relationship between people or functions.

A considerable number of assurance organizational approaches have been tried by industry over the past 50 years. Most started with some sort of quality organization reporting at a relatively low level within the company. The need and drive toward unification of the assurance disciplines have forced an evolution of assurance organizations. The major structures that have resulted are reviewed below.

4.3.1 "Degenerate" approach

This approach puts the assurance organizations within the organizations whose work is to be critiqued. The assurance functions are then frequently subjected to "smothering" and potentially insufficient management attention—and are therefore somewhat ineffective. An example is shown in Figure 4.1.

4.3.2 Staff assurance organization

One variation of the "degenerate" approach (and one which is certainly better) is having a top staff manager responsible for the assurance activities. If the top manager is strong enough and has some support, this method is effective even with the operating functions buried deep in the "guts" of the company as indicated in Figure 4.2.

The top manager can be functionally responsible for the assurance personnel, but not administratively responsible. Another method—more normally used with this type of organization—is to have this manager reporting the evaluation of the assurance personnel's findings to top management. In either case, the assurance effectiveness is highly dependent on the effectiveness of the top manager.

4.3.3 Engineering-oriented structure

In a company whose main forte is engineering research and development, a common assurance organizational structure has a form similar to that of Figure 4.3. This structure allows an independent assurance check of the research, design, and development activities. It is characterized by unimpeded access to top management and by system/

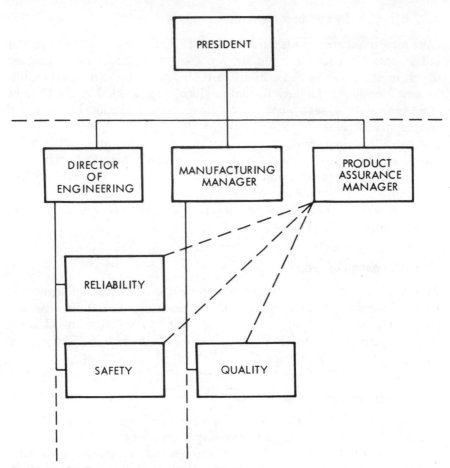

Figure 4.2 Staff assurance organization chart.

design oriented activities. There is usually no formal quality organization found in this structure—the quality functions are handled by one of the other assurance units.

4.3.4 Manufacturing-oriented approach

Corporations primarily involved in manufacturing activities are quite frequently organized with a quality/reliability assurance (QRA) function in parallel with the manufacturing function. This QRA organization sometimes contains all of the functions discussed in Chapter 2, but usually is heavily quality oriented as in Figure 4.4.

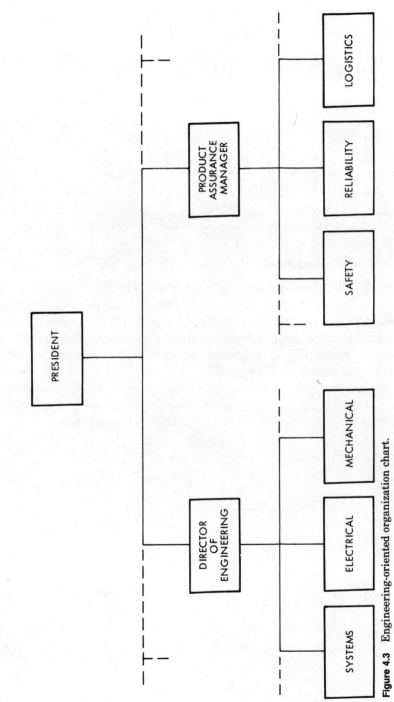

Figure 4.3 Engineering-oriented organization chart.

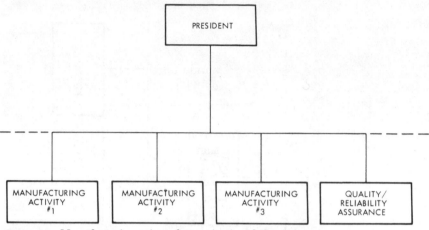

Figure 4.4 Manufacturing-oriented organization chart.

4.3.5 "Pool" organizations

Some companies have established a structure which sets up a "pool" or "homeroom" of assurance engineers. A small core of permanent personnel manage and accomplish the general organizational business. This almost invariably applies only to design assurance functions, i.e., usually to reliability or reliability/maintainability and not quality assurance as shown in Figure 4.5.

Personnel needed for a specific project are drawn for temporary assignment (duration of the project), then are returned to the "pool." This approach is not a favorite of professional assurance personnel.

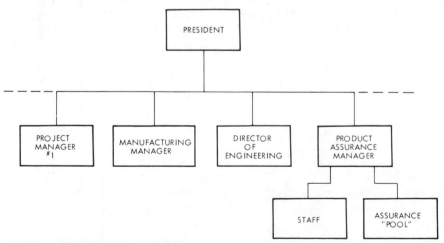

Figure 4.5 "Pool" organization chart.

Concern for personal development, lack of general assurance experience, difficulty in planning overall assurance activities, and the inevitable uncertainty are some of the reasons why this approach is difficult to implement effectively.

Many approaches have been tried and many combinations exist. All are probably effective—to some degree. The evolution of assurance organizational approaches has now progressed to the point where some generalizations about an "ideal" can be reviewed. The progress has occurred through integration of assurance disciplines and is characterized by lack of functional duplication. Like functions are now becoming one organization, e.g., reliability together with maintainability engineering, parts engineering together with incoming inspection of parts, etc. Specific characteristics of an ideal assurance organizational structure appear to be:

Ability to get adequate top management attention

Maintenance of a core of assurance professionals in the face of economic pressures

Ability to maintain a close working relationship with other functional groups

On a par structurally with engineering and manufacturing organizations

One possible organizational structure which embodies these characteristics and approaches the ideal (in a company that is both design- and manufacturing-oriented) is shown in Figure 4.6.

Obviously, any organizational approach must be integrated into the existing corporate structure without major upset. In very large corporations, one successful technique has been a small assurance staff at the corporate level which integrates the activities and sets the policies for the line assurance functions within each of the corporate divisions. This provides a single top-level point of contact and the ability to have somewhat different operating assurance organizations tailored to different divisional goals and overall structure. The small company can quite frequently accomplish its assurance goals by one single organization responsible for assurance functions—it might even be one person. However, this individual would still have to be able to obtain top management attention, maintain a close working relationship with other functional groups, etc., as indicated above.

An expansion of this "ideal" assurance organization, which embodies all the elements necessary for control as well as monitoring, is shown in Figure 4.7. A very important point shown here is the need for an *assurance data center*. Since one of the dominant functions of any assurance organization is the assimilation, analysis, and reporting of qual-

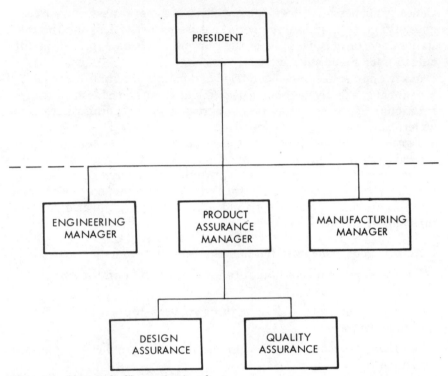

Figure 4.6 "Quasi-ideal" organization chart.

ity, reliability, safety, and field operation data, this piece of the structure is mandatory. Of course, in a very small company, this could amount to a portion of one person. In larger companies, the function could encompass complex computer systems and many personnel. In any case, the function is needed.

For functionally oriented assurance organizations in multiproject companies, the *assurance project management staff* has shown itself to be an efficient method of coordinating efforts. It is obviously not needed in a project-oriented assurance organization.

The existence of design and quality assurance follows the logic developed in Chapter 3 for assurance units associated with engineering and manufacturing efforts.

4.4 Staffing and Team Building

One problem—which has existed for a number of years—has slowed the growth of the assurance disciplines, both professionally and organizationally. Relatively few academic programs exist to train assur-

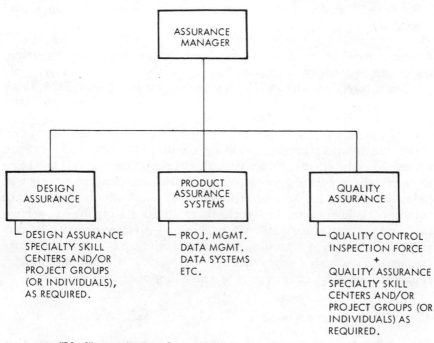

Figure 4.7 "Ideal" organization chart.

ance specialists or generalists. (Note: Chapter 14 reviews some of the recent trends and curriculum in academic training for assurance disciplines.) This problem has made it impossible to obtain trained personnel directly from college in the manner that design or manufacturing/industrial engineering functions do. Assurance personnel have historically been developed from line personnel by experienced assurance personnel—similar to the artisan approach. There are now quite a few state-of-the-art courses and industrial engineering curricula oriented toward quality control, but design assurance engineers (reliability, human factors, etc.) are still primarily "made" personnel.

The most successful approaches to staffing and building an organization appear to be:

Hire a committed professional as the top individual.

Hire a "core" group of specialists in the disciplines needed in the organization.

Bring interested personnel into the organization from manufacturing and design engineering.

Establish a new college hire program for "new blood."

Set specific assurance goals and objectives toward which the team will strive.

A transfusion of some personnel from design and manufacturing, trained in reliability, maintainability, quality, etc., usually prevents the "ivory tower" syndrome. Having been in these organizations, these individuals get more involved in assessing their old organizations' activities.

If the new assurance organization personnel are developed to understand the *total* assurance organization and its objectives, a team approach can be started. Quality assurance personnel should have some understanding of design assurance activities and vice versa. This will help in reducing intraorganizational conflicts. A group of assurance personnel understanding how each other's efforts are involved in achieving common assurance goals will be a team—given supportive leadership. One method of obtaining this understanding is "cross-training," i.e., letting assurance personnel actively work in the areas of safety, reliability, quality, etc., as part of a training program.

4.5 Motivation and Training

Many assurance personnel, after years of working in their specialty, develop a "second class citizen" feeling. The last section of this chapter describes some of the major reasons for this. Suffice to say that it is a real motivational problem, which is different, but perhaps no more severe than motivational problems in design engineering and manufacturing organizations. The establishment of an assurance team concept with common purposes will go a long way toward eliminating the problem and a well-planned internal training program will help even more. As mentioned previously, assurance management can set up a training program which gives each assurance specialist some insight into another's work. Quality personnel can spend some training time in maintainability and vice versa. This has the advantage of promoting team feelings, broadening personnel, and making specialists work more effectively.

In addition to this type of training, more formal educational courses now exist in a wide variety of forms. Programmed learning correspondence courses exist in reliability, quality control, statistics, understanding electronics, etc. Most colleges and universities offer many undergraduate and state-of-the-art courses directly associated with assurance disciplines. Assurance management should develop an atmosphere where this type of training is considered well worth the time. This is particularly true when companies pay all or a part of the tuition. Training (and sometimes certification) is also available from professional societies.

The Institute of Electrical and Electronics Engineers (IEEE) and the National Security Industrial Association (NSIA)—to name two—have subdivisions covering the assurance disciplines. The American Society for Quality Control (ASQC) is devoted entirely to segments of the assurance disciplines. Membership in these organizations makes some training immediately available. More importantly, the specialist gains the feeling of being part of a group of his or her professional peers. The symposia, workshops, and publications make available a wealth of assurance knowledge and practical problem-solving methods not available in textbooks. Assurance management can and should provide an atmosphere where professional participation is "part of the job." An absolute necessity for providing this atmosphere is the personal involvement of assurance management—providing the example by doing. Efforts should be made to interest assurance specialists in writing papers/articles for publication. In other words, the best long-range training approaches turn out to be significant professional motivation tools.

4.6 Special Purpose Groups

Why is there need for special-purpose assurance groups? Perhaps the best way to answer this is to ask another question—why have a grand jury, blue-ribbon committee, or crash investigation team? The creation of this type of group is generally valuable when there is a problem or breakdown in the assurance disciplines within a company. It is also quite valuable to concentrate the efforts of a few top-notch assurance experts on any high-priority problem. The bringing together of a small group of personnel with varying assurance backgrounds with short-term total involvement in a specific project frequently works wonders. This approach generally operates even more effectively if top specialist personnel from other parts of the organization are included. This type of "tiger team" does have drawbacks, however. Used too often, it tends to splinter the assurance organization—defeating the team concept that is so important in developing effectiveness.

4.7 Political Realities

It is a misstatement to say that top management understands the principles behind the assurance disciplines. The rising consumer organizations; the confusion between definitions for quality, reliability,dependability, etc.; the psychological name calling (e.g., "ilities," support groups, "toll gate" inspectors, cults); and the proliferation of assurance consulting firms all illustrate this. In addition, many businesses tend to consider the assurance work as overhead expense and—when facing severe economic pressure—cut back these functions first. This lack of understanding is a major problem to most assurance organizations. It makes "selling assurance" a continuing way of life for all assurance personnel. In contrast, chief engineers have no need to sell design engineering, they must merely sell their brand.

One of the main reasons for this problem is that it is difficult to relate assurance activities to dollars and cents. It is certainly true that savings result from an effective assurance approach, but quite frequently they are "cost avoidance" types of savings. Consider the inspector who detects an unrepairable process failure in making 1000 printed circuit cards before the parts are mounted. If the parts are mounted and then the defects are found, a gross labor and material loss will occur. How about the reliability engineer who, by analysis, determines that an equipment design will not pass a reliability demonstration test scheduled to occur at some future point? In each case, the assurance technique prevented a later cost problem. Too many managements will not recognize the value of this. In addition, management requires that functions like assurance justify their existence more heavily than those that produce saleable hardware.

All of this should convey a clear message—the assurance organization must sell itself. The sales effort will have to be continuous and based on cost savings to the largest extent possible. It will have to be based on performance and professionalism. And it *will* be a hard sell.

Lastly, some philosophy and historical perspective on assurance organizations. Twenty or so years ago, it was quite fashionable to discuss and actually attempt to implement the concept that design assurance requirements are part of a design engineer's job. This never really worked in a satisfactory manner. One possible very real reason for this is that there is *substantive work* involved in accomplishing design assurance tasks. This work, added to the activities and the schedule pressures already faced by a design engineer, overwhelms individual designers.

In contrast, the current trend of requiring manufacturing workers to be responsible for their workmanship adds little substantive *real work* to their jobs. It requires mostly motivation and training. The motivation and training approach has been demonstrated to be effective and is being actively employed by the more progressive companies.

WARNING: Parkinson's law begets dilution of product integrity achievement!

It should be obvious that assurance organizational growth for growth's sake will make selling product assurance an impossibility. It is probably true that assurance management must resist this even more than other operating organizations.

A rational organizational structure is necessary for effective management in all parts of business—no less so in the assurance disciplines. The particular organizational approach to be used for a specific company depends on a considerable number of variables. The variables will be . significantly affected by top company management as well as assurance personnel themselves. Development of an assurance team, somewhat independent of whatever structure exists, will have an overwhelming effect on the success of the assurance functions. The next chapter discusses how to effectively plan the activities of an assurance team.

5

Planning

*Forecasting is a difficult task, especially
when it is about the future.* An Irishman

5.1 Introduction

Anytime we philosophize on the subject of planning, Robert Burns'
famous observation that "The best-laid plans of mice and men gang
aft agley," comes to mind; and true it is. On the other hand, reflect on
how many more "slips twixt the cup and the lip" there might be if there
were no plans from which to stray. The ability to plan is humankind's
secret of dominion over all the other beasts that roam the earth. Whatever order is present in our daily lives is the result of formulating and
executing plans. Our daydreams, petty schemes, intrigues, formulation
of personal ethics, and expectations of immortality are all plans. True,
they often dwell more on the objectives than on how to achieve them.
Nevertheless, most of the important decisions in our lives have been
preplanned.

   Achieving the objective of a well-laid plan can be foiled only by sloppy
execution, or by an act of God. Business objectives are more often
successfully achieved than are our individual personal objectives
because the business environment forces *formal* planning. In our experience, the most striking evidence of the value of formal planning is a
comparison of U.S. armed forces training schools with public/private
institutions of learning. In the services every conceivable circumstance
must be *planned* for. Military technical training schools are *consistently
good*. College and university courses, as many students and graduates
will attest, are mostly dependent on the professor. Most are dull, some
are interesting, a very few are absolutely inspired. On balance, mili-

tary schools tend to be superior. How can this be? Military technical school instructors are relatively young and certainly lacking in experience and qualifications, as compared to college professors. The military instructor has two significant advantages: (1) the material to be presented, its order of presentation, and most of the possible reactions have been planned and prepared by highly qualified people, (2) a *lesson plan* for each session with the students must be prepared. College professors, on the other hand, do not have these advantages, or their constraints. Therefore, they have more time available to free-lance or to plan their lecture their way.

Consequently, rigid, formal planning in the field of education results in the most efficient transfer of knowledge. It is historically obvious that *addition* to humankind's knowledge most often originates, not in the rigidly planned instructional environment but among the "free thinkers." A good case can be made that this is so because invention and innovation are not among the *objectives* of the former environment. The determination of whether regimented planning is the natural enemy of invention is left to the interested student. Getting a job done is more efficient if it is well planned.

Planning is the process of establishing the what, why, when, where, who, and how of achieving some desired result. It is our method for shaping future events, limiting uncertainty, and avoiding surprises. When engaged in planning, it might help to recall a little prayer: "Give me the strength to change those things that I can change, the grace to accept those things that I cannot change, and the wisdom to know the difference."

Lack of the aforementioned wisdom leads one to waste energy against the proverbial stone wall or, on the other hand, to accept something as inevitable because "it's always been this way" or "they say it can't be done."

It is not by coincidence that the elements of planning include the "five Ws" of journalism. A good report is a comprehensive description of a past event, while a good plan is a comprehensive description of a future event. The "why" of planning is a particularly useful planning tool. Nevertheless, although the "why" is a powerful element of the planning process, it is a nonexecutable element. It is essential to the planning process but not to the resulting plan itself. Generally, if the what, when, where, how, and by whom something is to be done can be established, and all the why's are identified, an unassailable plan results. Given proper execution, the plan is invincible.

5.2 Requirements Analysis

Before planning any endeavor, a determination must be made of the limits of the ever-present constraints on our plans. First, what kinds of planning requirements are dictated by the nature of our business? This has already been covered, in a very large sense, in Chapter 3. Some rules were established for planning the structure of an effective product assurance organization as a function of the nature of our business. At this point, assume that there is an effective organization—such as that described in the preceding chapter. The problem now is to plan a modus operandi for this organization, and then to plan the effective execution of the various projects for which this organization is responsible.

The organization's modus operandi, when formalized, will be called the *operating plan*. The operating plan is long term, but sufficiently dynamic to keep abreast of changing business and technology. It sets forth objectives for improving the organization's performance and the means for achieving these objectives within a specific time frame. The product assurance operating plan should set forth objectives which are part of, or complementary to, overall company objectives. The resources available for pursuit of the objectives are likewise controlled by company policy and overall planning. Requirements analysis in preparation for the operating plan, therefore, consists of analyzing overall company goals and their relationship to the legitimate objectives of the product assurance function. For instance, suppose a company has established the following objectives for a given operating period:

1. Maximize profit
   a. Choose and pursue the most lucrative new markets and products which the company can satisfy.
   b. Minimize operating expenses.
2. Enhance competitive position
   a. Maintain and improve in existing markets and product lines.
   b. Build and maintain in the chosen new markets and product lines.
3. Build, maintain, and improve customer satisfaction.
4. Provide pleasant employment.
5. Build, maintain, and improve community relations.

Members of an astute product assurance organization should examine it in the light of these objectives, asking, "What can I do to help?" This question can then be broken down into a multitude of questions such as:

What are the candidate new markets and product lines?

Is there any advice regarding the risk/cost of providing product integrity in the new marketplaces which we might develop and offer?

Do we need any special preparation to enter new markets (analytical techniques, test equipment, inspection methods, etc.)?

Does the integrity of our present product lines compare favorably with that of our competitors?

Having compiled a comprehensive list of such questions and their answers, we have a good idea of *what* needs doing. A study of priorities associated with the company objectives and the resources available to our organization, leads to a determination of *what* we must plan by *when,* and some constraints on *how.* The rest of it is left up to us.

Unfortunately, there are usually even more constraints on our more common, work-a-day plans for various projects. Ideally, this would consist of a product assurance program plan for each program or project. This plan would embrace all product assurance activities required to ensure that the project results in a product of the required integrity. Given our "ideal" product assurance organization, this kind of project planning can be routine—even if some segments of the plan must sometimes be separable, as in "stand-alone" plans (to satisfy customer documentation requirements).

For a given project, the product assurance program plan must be in consonance with the operating plan and is also constrained by:

1. Customer requirements, both expressed (contractually) and implied

2. Product characteristics such as:
   a. Performance requirements
   b. Intended use
   c. Application environment
   d. Maintenance and support environment
   e. Production quantities and schedules

Product characteristics are, of course, often specified by the customer. The customer requirements above refer to other customer impositions on the manner in which the contractor is to design, develop, test, manufacture, document, and support the product.

5.3   Task Networks and Schedules

The real heart of what is to be done and how it is to be done is planned by defining work packages and writing task descriptions. The first step is to break the total job into manageable pieces. A manageable piece is a task that can be performed by a single organization, group, or individual; has easily defined content, input requirements, and output; can be accomplished in a relatively short time; and has measurable products or events that should occur frequently. It is obviously important that the sum of the work packages, or tasks, equals the overall job to be done.

The task description must concisely define the content, input, and output of the task. It is, in effect, a capsule statement of work. As such, it must leave no doubt as to what is to be done and minimum doubt about how it is to be done. It should not rely on reference to "standard operating procedure" unless that procedure is documented somewhere.

In concert with task definition and description, task interdependencies must be accounted for. One widely used technique for accounting for these interdependencies and their impact on cost and scheduling is PERT (Program Evaluation and Review Technique). Figures 5.1 through 5.3 are a series of example PERT charts which illustrate the steps in their development. The first chart illustrates the interdependencies of the tasks—which tasks require completion of which other tasks. The second chart incorporates schedule planning. Although the Gantt chart (see the schedule in the Product Assurance Program Planning Model, Appendix A) is a more readable visual display of schedule information, PERT scheduling is a more powerful analytic scheduling tool. It allows for a statistical accounting for the uncertainties of planned schedules, determination of critical paths, etc.—providing the information necessary to trade-offs in allocation of resources. In short, a schedule should be *developed* by means of some analytical technique such as PERT (or some operations research kind of technique) and *displayed* in a more simple visual form. For effective project schedule *control,* it is often necessary to return again to the operations-research type schedule analysis.

5.4   Cost Breakdown

The third PERT chart incorporates that parasitic element, cost. For statistical accounting of uncertainties in cost estimates, use the same method as for schedules. It should be noted that this should be done only for those elements of cost which are predominantly time-dependent (which should be minimized; level-of-effort cost estimating is a poor way to do business). The example chart doesn't tell the whole story concerning the level to which cost estimates should be broken down. It does illustrate that costs are to be broken down at least to the level of tasks to be controlled. In addition, when a physical product is involved, costs sometimes need to be further broken down into recurring and nonrecurring. Nonrecurring costs are typically those which accrue against the design and development of the product and do not recur for each copy of the product made. Recurring costs are associated with the physical product and are a function of the number of copies made. Recurring costs sometimes need to be broken down further, so that they can

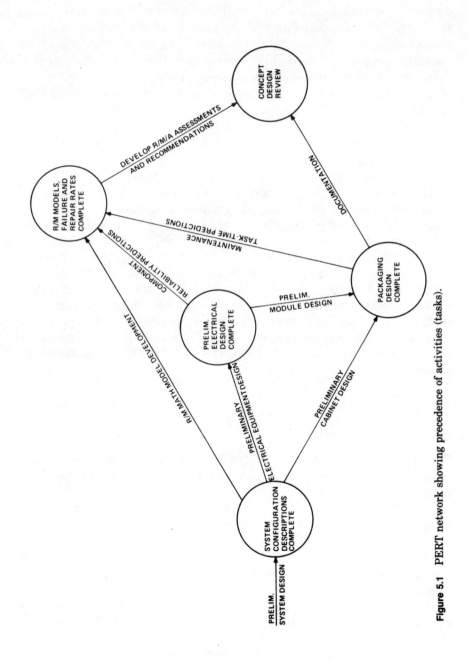

Figure 5.1 PERT network showing precedence of activities (tasks).

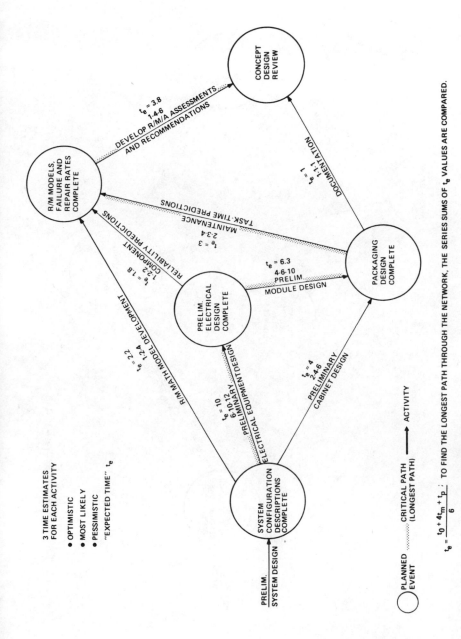

Figure 5.2 The network of "expected" times.

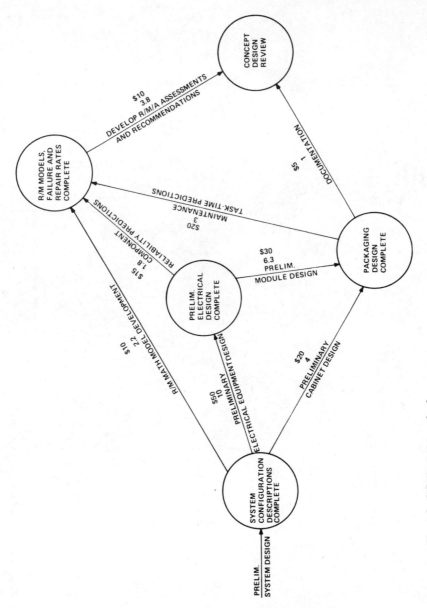

Figure 5.3 The schedule/cost network plan.

be associated with specific subdivisions of the product. In short, the level of cost breakdown depends on anticipating the needs of future planning and project control.

5.5 Planning and Trade-Offs

Cost, schedule, and risk form a terrible triumvirate. All plans contain risk—risk that technical objectives cannot be achieved, or that they cannot be achieved on schedule or at planned cost. Cost, schedule, and risk form a triangle in which modifying any one element causes changes to one or both of the other. For instance, costs can often be cut by reducing or deleting some task, thereby increasing the risk that technical objectives will not be met or that, in order to meet them, the schedule must be extended. Schedules can sometimes be shortened by planning for greater expenditures. Such trade-offs are a very important part of comprehensive planning. Trade-offs with respect to cost and schedule versus risk to product integrity are a particularly important aspect of product assurance planning, and they are discussed at some length, from time to time, in succeeding chapters.

Appendix A combines instruction with example for guidance in the development of a product assurance program plan, which is called here a product assurance program planning model. In structure, format, and in much of the wording, it provides the skeletal framework for a comprehensive plan for assuring the integrity of any product. In addition, it provides general guidance for tailoring the plan to a specific product, in terms of the nature of the product and its intended use.

5.6 Recapitulation

Formal planning is necessary to the success of any reasonably complex endeavor. It is particularly important in the quest for improving product integrity, since this quest requires the bringing together and concerted action of many disciplines—through the product's entire life cycle.

Planning need not be dry and dull. It can be creative—even interesting. For instance. if you are in the business of designing, developing, and manufacturing electronic equipment for the government, you could develop a completely "fleshed out" product assurance program plan around the skeleton provided by the planning model. If you use careful subparagraph structuring and wording (planning ahead), you will find yourself with a set of "modules" (paragraphs and subparagraphs), which, taken in the proper combinations, can be used to formulate most of the program plans required to assure reliability, maintainability, system safety, software quality assurance, etc.

In this chapter, the idea of project control has crept into the discussion quite frequently. There is no point to planning a project, unless the project performance is to be controlled to that plan. The following chapter treats effective project control, using well-laid plans as the standards against which to measure.

Controls

Man could direct his ways by plain reason.
SYDNEY SMITH

6.1 Introduction

Any company, whether large or small, has overall goals that it is striving to attain. These goals usually get translated into objectives that can also be imposed on a specific job. For example, a company may have a goal to improve its overall profits by 10 percent with an allocated objective of 15 percent to its "widget" business line. Along comes the XYZ widget job with its technical requirements, overlaid with cost and schedule constraints. How does a company make sure that its overall goals and objectives, as well as the requirements of its various jobs, are satisfied within given constraints? It doesn't just happen! There must be the effective application of a set of technical, cost, and schedule controls. These controls comprise a "kit" of tools which contribute to the attainment of the overall company goals and more specific objectives. As discussed in this chapter, these control tools are applied to ensure effective *assurance* efforts for any given job. Consequently, implementation of these controls by the assurance function contributes to achievement of the overall company goals and objectives.

In most respects, the basic control approach employed by the assurance functions on a specific job is similar in principle to those employed by other company organizations. Basically, any control approach should embrace a closed-loop system which identifies impending disaster before it occurs and provides proper and timely feedback for effective corrective action. The major differences in assurance control are (a) what is being controlled and (b) what serves as indicators of out-of-control situations.

What then is the approach to assurance technical, cost, and schedule control? Figure 6.1 presents the general approach in flowchart form; the steps involved are described below.

Establish the overall company goals and objectives. These goals and objectives are going to be directly or indirectly related to dollars—sales, production, profit, and cost.

Translate the company goals into overall objectives for a particular job. These job objectives are expressed in terms of technical considerations as well as cost and schedule. In order to do this, it is necessary to answer such questions as:

What are the company's policies toward design, manufacturing, and test activities (e.g., will failure reporting be initiated at all levels of product testing even though not required by the customer)?

What are the technical requirements?

What is the delivery schedule for hardware and documentation?

What is the target profit margin?

It is easy to see that some of these questions are answered by studying what the customer wants. On the other hand, some questions require that a decision be made by the producer.

Establish the assurance organizational goals and objectives. In a general sense, they are derived from the overall company goals and objectives. (In some cases, the organizations themselves have contributed to the establishment of the overall goals and objectives, particularly when policies are involved.) For a given job, the overall goals are categorized into objectives, such as technical requirements and cost/schedule constraints, for the assurance function.

Establish the implementation plans for achievement of the assurance organizational goals and objectives on the particular job. Depending on such considerations as the size of the company, its policies, and the documentation requirements of the customer, these plans could vary considerably in terms of depth, scope, and formality. They could be written as an informal internal memo or as a formal, full-blown program plan that gets delivered to the customer. In any case, there needs to be a plan of some type that tells what is going to be done, how, by whom, and when. Since *what* and *how* imply expenditures, plan implementation obviously requires dollars.

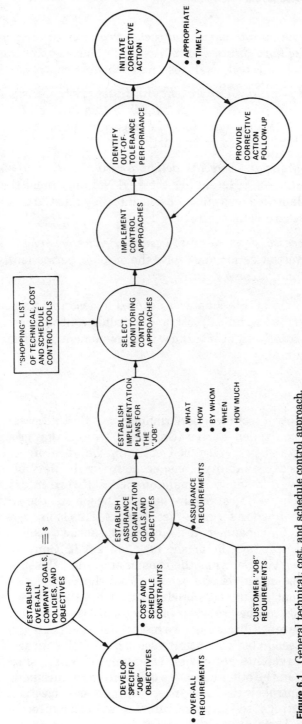

Figure 6.1 General technical, cost, and schedule control approach.

Select the monitoring control approaches to be applied to ensure accomplishment of the assurance job objectives.   These approaches require the use of certain technical, cost, and schedule control tools.

Implement the control approaches and identify the existence of out-of-tolerance performance (or potential) in technical, cost, and schedule areas.   Suitable indicators of undesirable performance must be developed to trigger corrective action.

Initiate appropriate corrective action to resolve and rectify out-of-tolerance conditions.   Herein lies the key! Not only must the problem areas be identified on a timely basis, but they must also be *resolved;* otherwise there is no control.

Provide necessary follow-up to ensure effectiveness of corrective action.   Corrective action must solve the problem permanently. It must go away and not come back.

The remainder of this chapter is devoted to describing a potpourri of commonly applied technical, cost, and schedule control approaches and to providing some insight into the way management gets visibility into the control function.

6.2 Technical

In order to control technical performance, it is first necessary to know what is to be controlled. Earlier (Chapter 2) we said that product integrity consists of a predetermined optimum balance of performance; aesthetic appeal; reliability; ease, economy, and safety of operation; ease, economy, and safety of maintenance; software interaction; and consistency. The technical requirements of the job and the product need to be clearly defined in terms of these factors. This is important to both the producer and consumer. If the consumer doesn't spell out what is wanted in the product, he or she may not get it. Conversely, if the producer doesn't know what the consumer wants, the producer may not meet those demands. Worse still, from the producer's viewpoint, overreaction can set in and a product that more than meets the customer's demands can be provided—at a higher cost.

For the assurance disciplines, we have noted that these requirements can be expressed in both quantitative and qualitative terms. The quantitative requirements are obviously easier to control since they are measurable. On the other hand, the qualitative requirements are more difficult to control since compliance with them is subjective (how good is good?). Control of the latter requirements is often directly related to the producer. How much can be spent to stay out of product liability suits, save the company's reputation, and minimize complaints? Just as compliance with these qualitative requirements is subjective, so is their control. Nevertheless, the requirements need to be defined in such terms that they can in fact be controlled. In many cases the producer will do the defining (Mrs. John Q. Public isn't going to establish the number of defects per 1000 dresses inspected—the dress manufacturer is).

Once the technical requirements are sufficiently defined, the approaches taken to show compliance with these requirements must also be defined. During the design phase, the producer performs certain tasks (e.g., software code reviews, design reviews, "paper" analyses, etc.) to estimate from the design whether the right solutions are being used to fulfill these requirements. During the build phase, verification (e.g., software validation and reliability demonstration tests and final inspections) continues to demonstrate compliance with these requirements. Finally, during the use phase, the user will determine whether the producer has done the job right and truly met the requirements. Having defined the means required to show compliance with the technical requirements, the producer can then establish and implement the appropriate technical controls.

Knowing what the technical requirements are and what approaches are to be implemented to show compliance, assurance personnel have various control tools (some of which are unique to the assurance functions and some of which are not) that they can employ. These tools form part of the following technical control approaches for control of: (a) requirements response, (b) design, (c) parts/materials, (d) design change and configuration, (e) subcontractor/supplier, (f) manufacturing, and (g) corrective action. Since these approaches and tools are discussed in more detail in later chapters, they are only briefly described in the section that follows.

6.2.1 Requirements response control

Control of product integrity requirements must start early with initial receipt of the customer's requirements. The company proposing to do the job needs to recognize the existence and importance of all requirements. Inability to recognize all requirements at the outset has at least two possible consequences:

1. The bidding company fails to respond fully to the assurance requirements. Thus, in the eyes of the customer, that company is nonresponsive and perhaps will lose the job.

2. The bidding company acquiesces to product integrity requirements with which it really cannot technically comply. Nevertheless, it wins the job because of a lower price. However, such a company could still

be a loser. Subsequent discovery of deficiencies could cost the "lucky" company both immediate dollars in redesign costs and reputation.

Therefore, there must be appropriate procedures to make sure that assurance requirements *and* risks are adequately recognized, evaluated, and addressed. After the job is in, the control techniques described below are applied to ensure compliance with the customer's product integrity requirements.

6.2.2 Design control

The "good" design of a product is obviously a prime prerequisite of product integrity. If the product is not designed properly for the way in which it is going to be used (this includes who, how, and where) it simply won't "fill the bill"—in spite of all the other assurance controls one might apply. The best time to influence the design is during the design phase, not during manufacturing, test, or use phases. Generally speaking, the further downstream from the design stage the product is, the more costly it is to change its design. Witness the costly recall programs automobile manufacturers have been forced to implement in order to correct design safety deficiencies. Thus, hardware, software, and system design reviews provide a cost-effective opportunity for doing the most toward achieving product integrity.

From an assurance viewpoint, design review provides the opportunity to critically evaluate the design from a product integrity slant and to bring to bear all of the analyses (e.g., reliability predictions) that lend insight into the product's integrity. To be effective, a company's design review program should be characterized by:

1. Design reviews conducted on all new designs and design changes
2. Design reviews performed at timely points in the design cycle (e.g., prior to release of drawings for fabrication and software code to "manufacturing")
3. Design reviews that are preplanned and scheduled
4. Design reviews that are documented

In short, each company needs to establish a clear-cut policy of what designs are to be reviewed, to what level, when, and by whom.

6.2.3 Parts/material control

The whole is only as good as its parts. When it comes to one of the important factors of product integrity—hardware reliability—this is no idle saying. This lesson was learned early in the U.S. missile programs when unsuccessful launchings were traced to a single failed part and/or faulty materials. Since then there has been considerable emphasis placed on assuring that parts and materials are of good quality. So much so, in fact, that parts/materials and how they are put together in the design and during fabrication have become the two most important factors affecting product integrity. Typical data from a wide range of products have shown that over 80 percent of the total anomalies experienced were due to faulty parts or workmanship. Is it any wonder then that parts/materials control has become an important aspect of technical assurance control? Implementation of parts/materials control serves to enhance the chances of acquiring the best parts and materials within the constraints of the budget. Effective parts/materials control includes such activities as:

1. Developing preferred/standard parts lists (give the designers a "shopping" list of proven parts)
2. Accumulating substantiating data before using nonstandard parts
3. Providing parts support to designers (e.g., parts application guidelines)
4. Developing parts/materials specifications to be imposed on suppliers
5. Employing source inspection at supplier facilities
6. Imposing parts/materials handling and traceability policies
7. Accomplishing parts/materials evaluation tests

6.2.4 Design change and configuration control

From an assurance viewpoint, it is particularly essential to keep track of hardware and software design and configuration changes. Often, these changes are made to enhance cost and producibility. However, there may be a detrimental effect on the assurance factors. Therefore, the impact of these changes on reliability, maintainability, quality, human factors, and safety must be assessed. In order for these changes to be assessed, the assurance functions must first know that the changes are being contemplated. Through a design and configuration change control system, employed by many companies, the assurance disciplines retain change review and signoff privileges and responsibilities. Thus, a measure of control is afforded for design and configuration changes.

6.2.5 Subcontractor and supplier control

Obviously, most companies cannot design and/or make all the items that they need in their end products. Yet, these companies still are responsible for ensuring that the total end product possesses the necessary degree of product integrity. Therefore, the overall assurance requirements for the end product must be translated into like requirements for the various subcontractors/suppliers. Appropriate steps must then be taken to ensure that the selected subcontractors/suppliers, whether they be for hardware or software, comply with the established requirements. To this end, appropriate control must be applied during both the subcontractor/supplier selection phase, and the procurement phase. This control includes:

1. *Surveys* to evaluate potential subcontractors/suppliers (e.g., are the procedures to be employed adequate to meet the assurance requirements?) and aid the final selection process

2. *Specification* of quantitative and qualitative assurance requirements and acceptance criteria (the latter criteria could include acceptance through analyses and/or tests)

3. Periodic *reporting* of subcontractor/supplier status toward achievement of assurance requirements

6.2.6 Manufacturing control

Although the design phase is important from a product assurance viewpoint, it's really the end product that counts. A hardware design can embody a high degree of *inherent* reliability, maintainability, and safety; yet these favorable characteristics can be degraded significantly during production. Furthermore, the stigma of poor quality can be introduced into the product, both hardware and software (e.g., media). To prevent and minimize such degradation, a wide range of manufacturing control activities are undertaken:

1. Control of critical items (e.g., through the implementation of special handling procedures)

2. Control of purchased items (as discussed above)

3. Control of in-house fabricated articles (e.g., through in-process and final inspections, as well as process controls)

4. Control of nonconforming material (e.g., by segregation)

5. Control of inspection, measuring, and test equipment (e.g., through scheduled calibrations)

6. Control of manufacturing and inspection personnel (e.g., through certification programs)

6.2.7   Corrective action control

It is one thing to identify problems, but another to fix them. Therefore, there must be a set of controls to ensure that corrective action does, in fact, happen—and happen effectively. You may ask yourself at this point, "Are we placing controls on controls?" In a sense—yes. For example, the implementation of manufacturing controls doesn't necessarily mean that there will be no quality problems—100 percent inspection is *not* synonymous with zero defects—defects can and do occur for one reason or another. Thus, corrective action control is important *and* necessary. Corrective action must also be verified to ensure effectiveness. The mere act of developing a fix is insufficient. It must be incorporated and proven effective. There are too many incidents in which a supposed fix (say changing a line of software code) causes another problem elsewhere. So the key to corrective action control, as shown in Figure 6.2, is providing the necessary follow-up to ensure that once a problem is identified, it is indeed resolved.

6.3   Cost and Schedule

Cost and schedule control starts early with the initial planning of assurance activities. To summarize from the previous chapter on planning, the planning phase requires:

1. A breakdown of all tasks to be performed and an organizational assignment for accomplishment of these tasks

2. A detailed set of task descriptions for work to be performed to meet the assurance objectives and requirements

3. A detailed schedule of time-phased activities related to these task descriptions

4. An internal "contract" defining either verbally or in writing what (tasks), when (schedule), by whom (performing organization), and for how much (budget) as part of the total assurance program.

In terms of control, there must be adequate and timely monitoring/ evaluation of assurance task performance against cost and schedule. Finally, assurance personnel can employ cost and schedule control approaches such as (a) evaluating the relative cost of assurance, (b) utilizing cost forecasts, (c) utilizing cost-to-complete estimates, (d) eval-

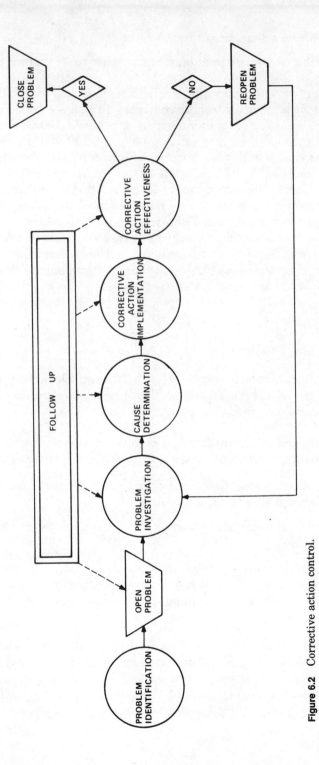

Figure 6.2 Corrective action control.

uating specific assurance job expenditures, (e) evaluating the cost of defects, (f) evaluating the cost of test failures, (g) evaluating the cost of maintenance service contracts, (h) evaluating the cost of warranties, (i) evaluating the cost of software maintenance, (j) utilizing milestone charts, (k) utilizing PERT, and (l) utilizing C/SPS (Cost/Schedule mance System). These approaches are described in the next section. Several words of caution should be noted before reviewing this material. The approaches presented here are among the most commonly used; they are not intended to be all inclusive. Nor do they need to be employed all together. Those reviewed are intended to serve more as a list from which to draw. It should be recognized that these control approaches are not unique to the assurance disciplines; some of them are used by others. All the approaches described here are intended to give advance warning of out-of-control cost/schedule situations; then when these situations do occur, to provide some insight into why. Finally, it should be noted that some of these schedules and cost control approaches (e.g., PERT/TIME, PERT/COST) were comprehensively developed in the previous chapter, and only a summary review is provided.

6.3.1 Evaluating the relative cost of assurance

One fundamental approach to cost control is to measure the relative cost of the overall assurance functions. Assurance organizational costs can be evaluated and compared against budget or a given standard. Admittedly the latter is difficult in light of the lack of universally agreed upon standards. Nevertheless, these assurance costs are important for cost control purposes. For example, it is important to know the ratio of inspection cost to production value, as well as the relationship between vendor inspection costs and the value and volume of purchases or number of shipments, and software maintenance costs as a percentage of total cost of software development. These costs can be tracked to see what portion of the overall company budget is related to the assurance functions, and whether these costs are increasing or decreasing. Any adverse trends can then be further analyzed to determine the reason why.

6.3.2 Utilizing cost forecasts

One of the more basic cost control approaches (not necessarily unique to the assurance functions) is the cost forecast. The forecast is used as an indicator of where the costs are headed and is usually done on a monthly basis. Each assurance task is reviewed in gross terms to estimate the costs to finish the job. These costs are time phased and forecast over the remaining portion of the contract. When added to the actual expenditures through that point in time at which the forecast is done, the cost forecast provides an up-to-date assessment of the cost of performing each task. Any deviation from budget can be quickly identified by comparing the sum of the actual expenditures and cost forecast results with the budget. Significant deviations from budget are then flagged for a more detailed review and the initiation of corrective action, if necessary.

6.3.3   Utilizing cost-to-complete estimates

A more detailed look at where the assurance organization is heading from a cost viewpoint is through the use of a cost-to-complete estimate. If the cost forecast identifies that projected costs are outside acceptable limits, then it may be necessary to perform an in-depth examination of the work remaining and estimate the costs related to task completion. Instead of examining the estimated costs through the remainder of the contract in gross terms, a miniproposal is prepared. Each assurance task is reviewed in terms of subtasks, and a rationale for doing each subtask is developed. For example, the cost forecast may simply indicate that the reliability task will result in a total expenditure of $100,000, versus a budget of $90,000. On the other hand, the cost-to-complete will take the overall reliability task and provide estimates for each of its subtasks. Therefore, the cost-to-complete will provide insight into the fact that (a) a comprehensive reliability prediction update has to be accomplished, (b) the failure modes analysis has yet to be started, (c) design review support has yet to be provided. In the cost-to-complete estimate, a detailed description would be provided for each of these subtasks so that the cost growth can be clearly understood. Thus, in our example, the design review support subtask could be expanded to provide details for design review preparation and attendance time, time to work action items, and travel dollars.

6.3.4   Evaluating specific assurance job expenditures

Another cost control approach is to review the expenditures on a given job and examine the cost breakdown of the assurance operations. In its simplest form, this approach can be limited to direct labor hour expenditures. These summaries can be issued on a weekly basis and used to track weekly, monthly, cumulative, final, and budget-authorized expenditures. A somewhat more expanded version of this technique is to incorporate other cost data. These data could include overhead, travel costs, agency help charges, computer rental, and other incurred costs as related to assurance operations on the job being monitored. Obviously, comparisons between actual expenditures and forecast expenditures, together with a look at the budget, can be used to identify potentially out-of-control situations.

6.3.5 Evaluating the cost of defects

With the notable exception of the approach of evaluating relative assurance costs, the previous cost control approaches were rather general. Nonassurance organizations could also apply the same approaches in principle. One specific measure relating to the assurance functions, however, is the cost of defects. This cost measure of product quality includes tangible costs such as (a) scrapped materials, (b) labor and burden on scrapped materials, (c) repair costs on salvaged materials, (d) excess inspection costs, (e) defect investigation costs, (f) customer complaint costs, and (g) quality guarantee costs. As inputs to quality cost control, these costs can be treated collectively or individually. They can also be further subdivided to provide keener insight into the major contributors within a factor. For example, customer complaint costs could be subdivided into claims, transportation, and investigation costs. From a control viewpoint, these costs can be presented in various tabular and graphical forms (e.g., bar charts) and compared with preestablished standards or past history. Figure 6.3 provides an example of a trend chart for customer complaint costs. The increasing trend in complaint costs provides an indication that all is not going well—action must be taken to establish the causes of customer complaints and to bring these costs back into line.

6.3.6 Evaluating the cost of test failures

Another more specific measure relating to the assurance functions is the cost of test failures. Although this cost can be broadly tied into the reliability and quality of the product, the reasons for test failures can be attributed to a variety of reasons (e.g, documentation or diagnostics errors). The cost of test failures includes cost elements similar to those related to the cost of defects (retest, failure investigation, and throwaway costs, for example). Moreover, this cost involves expenditures from various company organizations, both assurance- and nonassurance-related. For example, an investigation could be launched to establish the cause and corrective action necessary for a design-related failure problem; obviously, the design group would be one of the organizations involved in this investigation. A similar approach to that used in controlling defect costs would be employed to monitor and control test failure costs.

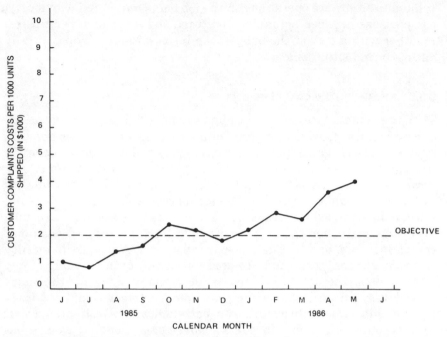

Figure 6-3   ABC Company customer complaint cost trend.

6.3.7   Evaluating the cost of maintenance service contracts

An important cost control approach, particularly in the commercial sector (for example in the computer and appliance industries), is the cost of maintenance service. Maintenance service contracts are written based on, among other considerations, assurance-related factors. Some of these key factors include an estimate of the number and kinds of failures or service calls, and a projection of time to perform maintenance, both on-site and back at the repair "shop." If the design is inherently unreliable or not easily maintainable, or the hardware has an excess number of manufacturing defects, or the software has considerable residual "bugs," the cost to the lessor of the contracts will begin to escalate. The actual cost for service will exceed the price spelled out in the maintenance service contract. Thus, actual service contract costs can be tracked and compared against the contract price, and deviations investigated so that appropriate corrective action can be taken. Obviously, the nature of the corrective action pursued will depend on the cost drivers, whether they be due to one of the assurance factors (design unreliability) or by some other factor (maybe the profit margin was set too high).

6.3.8   Evaluating the cost of warranties

Warranty costs are another area of good cost control. At one time, warranty costs were a concern solely on commercial products. Now, more and more warranties are being contractually imposed on military products as well. A "generous" warranty can go a long way toward attracting product buyers. However, it can also contribute significantly to unprofitability if there are extensive product problems which eat into the dollars allocated for warranty costs. Once again, the assurance factors are the key to keeping these costs down. The product must be reliable, free of defects, and safe to use. As in previous cost control approaches, actual warranty costs can be tracked and compared against allocations, and corrective action instituted as necessary, depending on the cause(s) of escalating warranty costs.

6.3.9   Utilizing milestone charts

So far we have discussed cost control approaches. How about schedule control? A simple way to track assurance performance versus schedule is through the use of any one of a variety of milestone charts. Since such charts usually form a part of an implementation plan (e.g., system safety program plan), they are usually readily available. If not, they can be easily developed. Failure to meet milestones, or changes in milestones for reasons other than nonperformance, can easily be noted on the chart. Similarly, a missed milestone can be readily flagged for corrective action. Figure 6.4 provides an example of a typical milestone chart. As noted in this chart, the start of the preliminary hazard analysis has "slipped" and been rescheduled. Such slippage, once flagged, can be evaluated for impact on the rest of the program.

6.3.10   Utilizing PERT

In addition to approaches which apply solely to either cost or schedule, there are also those that can be adapted to both. As you may recall, the PERT/TIME and PERT/COST networks provide a logical representation of the interrelationships between all program tasks and related costs. Their primary application is in major program efforts. Since the customer usually doesn't require it (because of the cost involved), PERT is rarely applied to small jobs or by small companies. As a control tool, PERT provides a way of controlling cost, progress, and change. Cost control is achieved through a continual comparison between actual and planned costs. Progress control is accomplished by reviewing actual against scheduled progress. Change control is brought about through a review and evaluation of changes (e.g., engineering changes) which occur during the execution of the project. Another control area avail-

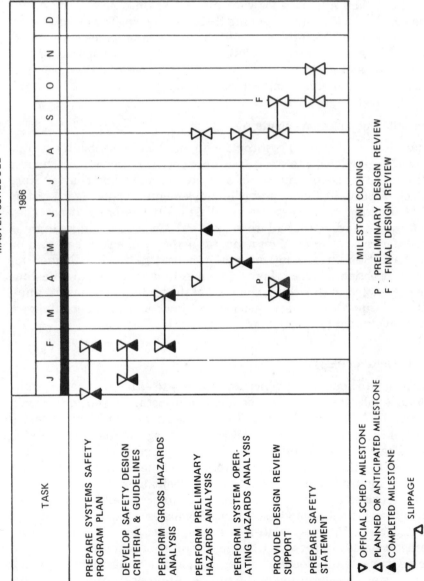

Figure 6.4 Widget systems safety program.

able through PERT is assessment of the adequacy of work being done on the project. Work adequacy control can be obtained by a continual review and evaluation into the effectiveness of work being performed.

6.3.11   Utilizing cost/schedule performance system (C/SPS)

Another technique which provides a most sophisticated and combined approach to cost and schedule, as well as technical performance control, is C/SPS. Control is accomplished through a regular review of detailed cost reports and calculated schedule/cost indices to identify out-of-tolerance cost/schedule performance. Once identified, the reasons and corrective action for such performance must be explained in variance analysis reports. As in the case of PERT, C/SPS is usually only implemented by large companies—and then only on major Department of Defense programs when contractually required.

6.4   Management Visibility

It is obvious that the technical, cost, and schedule controls described in the previous sections yield meaningful data. This data can be used to provide management visibility so that appropriate corrective action can be taken. Data concerning design, hardware, software, manufacturing, test, and use performance can be used to provide visibility into technical compliance. Similarly, data resulting from schedule and cost systems can be used to provide visibility into schedule/cost compliance.

Regardless of the management level, both formal and informal approaches are employed to obtain visibility of the presence or lack of assurance control. The media can take the form of written or oral communications. Whatever the approach taken however, there *is* duplication—personnel from different management levels sometimes do read the same reports and attend the same meetings.

The degree of visibility, in terms of scope and depth, that management personnel have in assurance matters depends on where and what they do in the overall company organization. At one end of the spectrum are the lower-level management personnel who are party to the details of assurance progress and problems. At the other extreme are upper-level management personnel who have the overview of the assurance situation. Somewhere in between these two levels are those management personnel who see a great deal of the details, while at the same time getting some of the overview.

Since lower-level management personnel are the closest to the everyday assurance progress and problems, they should have the best insight into what is happening in their particular areas. They have the opportunity to communicate, daily if necessary, with their subordinates in order to understand precisely how things are going. Their communication lines can consist of firsthand oral reviews of project performance and of various technical/schedule/cost reports.

Middle-level management personnel can and do rely on similar types of communications lines. Oral reviews can be conducted with key project personnel and/or the first-line supervisor. These management levels also may have the benefit of higher management documentation, briefings, and meetings with project personnel that are not available to lower-level personnel. Thus middle-management personnel are in a unique position in terms of management visibility—they can obtain an overview, or be close to the details of the assurance problems, if need be.

As a rule, higher-level management personnel do not become intimately involved in the details of everyday progress and problems. They need to have an overview into the situation. However, on critical projects or programs (e.g., those having existing or potential contract noncompliance and/or overruns), it may be necessary to have an in-depth review. Periodic project reviews can be intensified and extended to provide more details. Similarly, special project summary reports can be directed to track technical, schedule, and cost progress.

In summary, implementation of technical, schedule, and cost controls provide needed data to all levels of management: (a) to identify existing and potential risks and problems, (b) to serve as a basis for initiating corrective action, (c) to determine possible areas for trade-off, and (d) to use as an input to decision-making relative to establishing technical/cost/schedule priorities. The next chapter describes the types of documentation used to provide this and other assurance data.

Documentation

Communication is Power
SENATOR J. W. FULBRIGHT

7.1 Information Requirements

In this age of instant communication, both verbal and machine generated, the thought must cross everyone's mind, "Why all this paperwork?" Of course the primary purpose *is* communication, but there are
other reasons, both rational and irrational. Many jobs and organizations exist as a consequence of nothing more than the huge mounds of
documentation generated by industry each day. It is evident that self-
interest plays a role in our increasing documentation jungle. Some
people just like to see their name on a report with reasonable frequency. Others generate memo after memo to make sure that they are
covered. The examples of documentation misuse would fill a book.
Again, the main purpose behind recording anything on paper, in a
computer, on magnetic tape, on video tape, and so on, is communication—either in the present or in the future. Present documented communications span a range from memo through video tape and electronic
storage/retrieval. Documentation, for our purposes, means permanent
or semipermanent recording of information in or on some media. A telephone conversation is not documented communication.

One of the most significant reasons for paperwork is efficiency. Imagine trying to telephone 30 people to acquaint them with the contents
of a 50-page report. Not only would there be difficulty in getting an
identical message to each of the 30 recipients but no record would
remain of what had been said. This is an extremely important facet of
documentation. It remains as a reference baseline for future activities,

providing a permanent record of accomplishments, data, and status. One example of the inestimable value of product assurance documentation would be its successful use in a product liability action to demonstrate that proper assurance activities had taken place.

In a modern complex operation of a company, operational information is needed for all levels of personnel. The need for operational facts is no less severe in the assurance field. The manufacturing organization needs data showing number and type of defects, general quality levels achieved, etc. Design engineers need data on the reliability of parts and systems, maintainability goal achievement, and safety analyses. Marketing demands usable assurance sales pitches—MTBFs (mean-time-between-failures) of equipment, life data, environmental capability data, and the like. Customers (particularly governmental customers) now write very complex and expensive assurance documentation requirements into their contracts to satisfy their information needs.

All of this points out that documentation is one of the most important tasks that any assurance function undertakes. Assurance information is needed by every part of a company in some form or another. The defect information provided to manufacturing is also needed in part by design engineering to solve design-related problems.

The same defect data is summarized, grouped, and properly presented in reports to upper management to indicate the general quality and design "health" of the organization. The defect information is often used by accounting to maintain an analytical trend review of scrap and rework costs as a percentage of manufacturing material dollars. Defect information, categorized by vendors, is used to take corrective action with particularly unresponsive suppliers. Specific, by application, failure data is used by design engineering to identify longer term trend failures resulting from marginal design application and tolerancing. Of course, in the overall sense, management uses assurance data to measure the health of the company. Profits generally fall as defective material increases; increases in field failures can foretell a maintenance and warranty disaster; and increased human-related defects in production can even signal motivational or morale problems in the workforce. If the task of documenting assurance-related performance is done well, it goes a long way toward bridging any communication gap that exists between the somewhat unrelated parts of a company.

7.2 Communication

Assurance information should be aimed at documenting the performance of manufacturing or design engineering organizations in accordance with specific contract and company assurance requirements. The information must be aimed at showing that "all's well" or that some corrective action is needed to meet a specific requirement. It must be remembered that the assurance disciplines are primarily "see to it" rather than "do it" activities. This means that these corrective actions will generally be accomplished by some design or manufacturing function. In order to accomplish these actions, these other functions must have trends, analyses, and other data communicated to them. The way in which these assurance data are prepared, documented, and presented, determines whether "communication" will be achieved.

Assurance activities are almost always faced with large amounts of detailed data, the essence of which must be communicated to the remainder of their organization. These data must be analyzed, summarized, and presented in a concise understandable way (who is *really* interested in reading long narratives about one's mistakes?). Experience shows that visual displays of assurance data similar to Figure 7.1 are really used and understood. These can be hand prepared or done by a computer as in this example. The worst way to present data for most people's consumption is in large piles of computer output. Most of these piles of paper live only long enough to reach the "round file." Graphs, histograms, and "pie" charts are examples of visual displays well suited to documenting assurance information.

Assurance information prepared for upper management consumption will be more condensed than that distributed to personnel responsible for solving any detailed problems pointed out by the data. If each major assurance area can be reported—visually if possible—on a single page, chances of its use and retention will increase considerably. Reporting of this information to top management is generally best accomplished on an exception basis. This is sometimes dictated by policy and practice. Figure 7.2 shows an example of a single-page management report on the overall quality "health" of a company. Note the use of "up" as a positive sign (normally psychologically true), simple legends and scales, and the uncomplicated visual presentation of the data. It sometimes helps to be a psychologist!

A major advantage of a well-developed assurance reporting system is its sales value. As the Political Realities section in Chapter 4 pointed out, it is very valuable to the assurance organization to continue selling its capabilities at every opportunity. Daily, weekly, and other periodic professionally designed assurance reporting pays for itself over and over in public relations value. It pays to advertise! However, it should be obvious that this aim should generally be secondary to positively oriented problem solving and reporting.

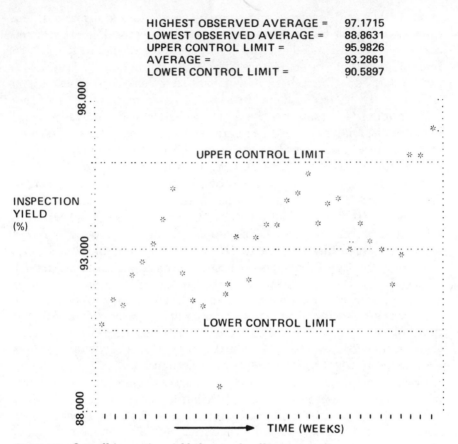

Figure 7-1　Overall inspection yield (four week rolling average).

7.3　Records

Reports often automatically turn into records, but the reverse is seldom true. One of the main tasks of all the assurance disciplines is to maintain a record system of detailed assurance data. Examples of these detailed data to be accumulated are:

1. Quality—inspection yield, numbers and types of defects, and process control data

2. Reliability—number and type of failures, operating hours, life data, and failure rates

3. Safety—major hazards, field safety problems, and hazard labeling

4. Maintainability—repair and down times and special tool requirements

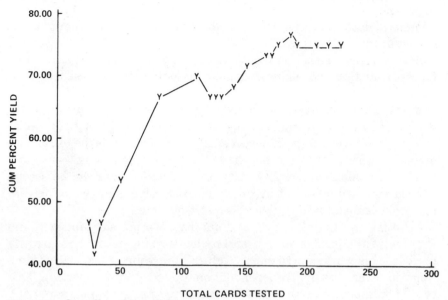

Figure 7-2   Overall card test yield.

This data should be organized and filed to facilitate easy retrieval at some indeterminant future time (the ASQC has developed an indexing system which can be helpful in making this possible). In some companies, it may be necessary to establish files by project, product line, transistor type, or some other part or equipment method. This can be as simple as putting reports, computer runs, punched cards, or whatever into file drawers. Increasing usage and simplification of microfilming and allied techniques has established economical systems for small-volume filing and visual retrieval. Large-volume filing and sorting generally is accomplished in computer memory. The filing technique will be largely dictated by record usage frequency and whether "hard" copies are frequently needed.

It should be obvious that there are considerable potential costs associated with any filing system. For any reasonable amount of data periodically received, a cost trade-off analysis should be done to ensure that the least-cost storage system is being used. Probably the most flexible filing system in use is a computer database, exercisable interactively in an almost infinite number of ways. The least flexible system is the hard-copy filing system (e.g., file cabinets). Filing and retrieval costs generally increase with flexibility. In any case, the value of filed historical records can be enormous. Few product liability suits can be successfully defended without a large number of records and a considerable amount of documentation that proves conscientious attention to all of the assurance disciplines.

Accumulation of assurance data makes it possible to accomplish many types of statistical and engineering analyses for unearthing design or quality problems not apparent in day-to-day activities. Sporadic failures of an identical component used in many different circuits over a number of weeks can be found. Workmanship problems relating to many assemblies can be pinpointed to a specific individual. More common is the use of data over longer periods of time, showing declining or increasing trends in a vendor's supplied part quality. Analysis of detailed data can be accomplished that will point out necessary circuit design changes to increase test yield against an inflexible parameter minimum.

Field failure, maintenance, and quality data properly indexed and eventually analyzed is quite useful in determining economic warranty periods. This works particularly well in high-volume product production. However, it is highly dependent on the accuracy of reporting. This type of data is also analyzed to determine maintenance personnel and facility requirements.

Last and certainly not least in value, indexed historical record files can be used for preventing "reinvention of the wheel." There are instances—especially in larger assurance organizations—where the same quality problem is investigated, analyzed, and solved every 2 or 3 years. Complex reliability analyses, useful in the general sense, are sometimes done a number of times (e.g., redundancy equation development, a "new" mathematical approach to reliability requirement allocation). In some companies, reliability analyses of the same power supply or computer are done over again on a different project. The economic value in preventing these types of activities is obviously significant.

The real problem with all of the preceding is that no one can gaze into a crystal ball and know that specific detailed data will be needed or be valuable in the future. This results in a large amount of detailed data being stored on the chance that it will someday be useful. Quite frequently the data does turn out to be valuable in some unforeseen way, but a continuous problem will exist in justifying the costs of data storage.

7.4 Information Systems

With the ever-expanding need and desire for more and better assurance data have come many attempts to simplify its gathering, analysis, reporting, and storage. Initially, since most assurance functions were organizationally separate, a wide variety of data reporting systems came into being. Reliability and maintainability functions established systems for gathering failure and repair data. Different systems were established to record defect and process control data. The factors described in Chapter 3 began to exert pressure, and organizational combinations began to appear. This was happening as both industrial and governmental information needs were mushrooming. The inevitable result is the present trend toward establishing common assurance data systems. Similar information needs in other parts of the organization (production control data, procurement data, engineering data, etc.) were forcing common data system developments also. These developments have created "information systems"; i.e., a common information base from which a variety of reports can be developed concerning different functions. The pressure to centralize *all* of a company's information needs from a planning, analysis, and implementation viewpoint has created yet more organizational growth. These functions are generally called something like "management information systems" (MIS). Presently, considerable organizational and philosophical furor exists as to how far this function goes in a given organization. Suffice to say that the centralization of data systems is where the action is.

Modern assurance information systems are established to gather data on incoming inspection, in-process inspection and test, in-process and field failures, and repair/maintenance actions. These data are stored in a common base, accessible to any of the assurance functions for analysis and reporting. The magnitude of these data generally dictate whether computer data processing techniques are necessary.

Since the development of a data processing system is a somewhat specialized effort, it should begin with a well-designed plan. The following list will provide a general idea of the key elements:

1. Develop clear and specific system specifications. (Computers solve only well-structured problems!)

2. Make the functions automatic. (Automated procedures alone will not suffice.)

3. Keep it simple—but integrated. (Use other databases while holding complexity down.)

4. Employ a modular system design. (Problems of overcomplexity will be reduced.)

5. Do not depend on informal data input. (No one talks to a computer yet!)

6. Provide for flexibility. (Build in auxiliary functions.)

7. Anticipate "feedback"-induced system changes. (This will avoid "topsy"-like growth.)

8. Make the end users of the data part of the development. (Why else
   have the system?)

If these elements are considered and a team of the users and data
processing specialists design the system, a useful and efficient system
should result.

In large companies with their own internal computer facilities, equipment and techniques are generally not a problem. Smaller companies
can, however, have their own assurance information systems relatively
cheaply by using any of a number of "time share" computer companies. Systems design support and standard programs from these sources
are quite useful. They have the feature of being inherently flexible,
frequently producing reports that need no "prettying up." Since these
types of systems are particularly user-oriented, many valuable built-
in features are already standard features—no special programs needed. The user languages are all extremely easy to learn and generally
free instruction sessions are maintained. This has allowed managers
and administrators to begin using secretarial and clerical personnel
on a much broader basis—making both happier. While information
systems of the sort described generally solve the problems of storage
and retrieval, care must be taken to prevent uneconomical system
growth. Usually a single knowledgeable individual can maintain cost
control of a time-shared system.

Another alternative for assurance information systems for smaller
companies is the personal computer (PC). The existence of a wide selection of very powerful PC database systems and PCs with significant
internal memory and storage makes this alternative not only possible
but relatively easy. The rapid development of PC networks and connection of PCs to much larger computers makes the PC approach to assurance information systems a cost-effective, flexible one for any size
company. The resultant ability of an analyst to download only data of
interest and manipulate it using the most commonplace PC spread-
sheet and graphics software yields very significant advantages in schedule and personnel.

Industry is already involved in completely automatic assurance information systems at this point. It is quite easy to imagine (an increasing
number of cases already exist) all assurance personnel typing or read-
ing-in results of tests, failure analyses, inspections, and investigations
as the events occur. All of the data would go to central data storage
where it would be categorized, sorted, analyzed, and flexibly stored.
Management could then interrogate the system—in simple English—
and receive specific answers to specific questions displayed on a screen,
printed, or vocalized. This would avoid the use of reams of paper, considerable manual analytic efforts, distribution problems, and many addi-

tional costs. It also solves the main problem that presently exists: the ultimate user is a number of minor steps away from the needed information, so he or she must explain what is needed to a number of intermediaries, with the resulting loss in resolution and definitions.

8

Design

*The more alternatives, the more difficult
the choice.*   ABBE D'ALLAINVAL

8.1 Introduction

Some of the more enticing and productive opportunities for truly useful
work in product assurance are those of influencing the design of a product. We have defined this segment of product assurance as design assurance. Design assurance is the companion to product design, as quality
assurance is to product manufacture. We will attempt to distill the
essence of design-influencing power out of the total "bag of design assurance tricks"—developing a set of rules for effective selection and application of the more important design guidance and review techniques.
The primary concern of design assurance is preventing or correcting
those design errors that lead to poor product integrity. If design errors
can be prevented—fine; otherwise, the earlier they can be detected and
corrected, the better. The concern here is with their detection, prevention, and correction before release to manufacturing.

Much has been written about the "how" of the basic design assurance techniques (Chapter 9, for instance). The concern here is with the
"when," "under what conditions and in what combination," or indeed
"if" these techniques should be applied. The "how" is addressed only in
the event of disagreement with common practice or where some favorite shortcut, caution, or trick for improving effectiveness or efficiency
is described. The quality of the craftsperson's work is highly dependent on the ability to select the right tool for the job at hand. So it is
with the design assurance engineer. Knowing *why* a certain tool or set
of tools is proper for a given situation is very helpful. In fact, this knowl-

edge can replace a good portion of the hard experience ordinarily required to become proficient at selecting the right technique. Above all, the selected techniques must be pertinent to the design problem at hand and to the current stage of solution of that design problem. They must have potential payoff commensurate with the required investment. The "design-active" end of a technique (reported results, recommendations, etc.) must be provocative. One must get the attention of someone who is in a position to make it happen. In this chapter, design assurance activities are treated in a more-or-less chronological order. Thumbnail case histories of dearly bought experiences in design assurance are described. These case histories will illustrate the truth of what is presented. If the most common and troublesome design errors are exemplified, then each case should evoke memories of similar events. The readers can then verify our advice for effective design influence— by their own experience.

8.2 Technical Requirements Analysis

Every new product, from the simplest device to the most complex system, begins as an idea. Likewise, every significant advance in designed-in product integrity begins with an idea. One of the more important ideas of design assurance is that it is not so much the absolute value of individual integrity parameters that govern the "goodness" of a product, but rather the proper *combination* of integrity parameters, together with the parameters of performance, capability, cost, and sometimes aesthetic appeal. Partial realization of this idea is attained by using the techniques of system effectiveness, integrated logistics support, and life cycle cost analysis. Full realization of this concept is achieved only by developing the optimum set of goals for product characteristics.

Development of the ultimate methodology for determining the optimum product characteristic goals is not up to product assurance. Even the practice of manipulating the existing techniques of systems effectiveness, integrated logistics support, and life cycle cost analysis is not the proper province of product assurance, except by default. These techniques are branches of systems engineering which embrace *all* engineering disciplines.

Nevertheless, the assurance scientist must participate in these ideas and the design assurance specialist, in particular, must shoulder his or her share of responsibility for determining the proper product requirements. Once some understandable bounds have been determined for product performance, cost, and aesthetic characteristics, and for product integrity characteristics, the assurance specialist must be prepared to assist in determining optimum bounds on the individual product integrity characteristics. In short, the question must be asked: "What kind of integrity is desirable for this product?" To find the answer to this question, such questions as the following must be answered:

Is the product durable or consumable?

Is product operation attended or unattended?

Can it be repaired when it fails? With or without interrupting product operation?

What will it cost the user for each incident, or each minute or hour or day that the product won't work? What are the consequences of product failure or inoperability?

Once the answers to an exhaustive set of such questions which are pertinent to the product development at hand are secured, the makings of systems effectiveness and life cycle cost models exist, tailored to this product. These answers, properly quantified, can be used with the models until a set of product integrity requirements are developed.

A usable set of product integrity requirements must be clear and unambiguous, preferably quantitative, and somehow measurable. The job of design assurance can be pursued; it consists mostly of avoiding the mistakes we have made in the past.

8.3   Perennial Problems

Human nature being what it is, 80 percent of all problems are caused by 20 percent of the population. This is just as true for "machine" designs as it is for the body politic. Over the years, a few classes of design errors have been found which cause a surprising amount of grief. Notable are:

Component misspecification

Part misapplications

Design oversight

8.3.1   Component misspecification

Any component of a product that is purchased, rather than "built" by the product designer is suspect—if for no other reason than the communication problem. It is very difficult to determine exactly what one's requirements are and even more difficult to express these requirements in an unambiguous manner.

Incomplete specifications. The more important parameters of a part, or component, are generally accounted for by specification. The trouble is we often tend to judge what is important by habit and tradition, rather than by the requirements of our present application. There is a tendency to overlook or ignore parasitic or transient possibilities. The following sequence of events generally defines the problem:

1. Determine the gross requirements for a part or component.
2. Select a likely candidate from a particular vendor.
3. Verify proper operation in a breadboard test and/or design around this particular component.
4. Write our specification around the vendor's specification for this part.

The component may then be purchased from any number of vendors. It may meet the specification but fail to perform properly in our application. Even if we buy only from the original vendor—if the original process is changed, the component may not work in our application, even though it continues to meet the requirements of the specification; this failure is most likely due to undefined, incompletely specified part parameters.

Unenforceable specifications. A parameter of parts and components which is frequently not controlled by specification is reliability. The specification that "The 6N444 Spritzistor shall have a failure rate not to exceed 0.001% per thousand hours" is an absurdity unless some provision is made to verify that this requirement is met. This sort of thing not only does nothing for product reliability, it gives the reliability discipline a bad name. There are those who insist that such phrases be included in every specification issued on a project. A similar situation arises with respect to attempts to control maintainability of components. Imagine a power supply vendor's confusion at being required to provide a power supply module having a mean-time-to-repair (MTTR) of 15 min, for organizational level repair, when the component is a replaceable unit in your system. "Quick-disconnect" connectors, malfunction alarms, etc., can be provided but there can be *no* guarantee of a 15-min MTTR for that power supply in your system. It is therefore necessary to specify the predetermined system developed specific power supply requirements which will result in satisfying the overall system requirements (e.g., quick-disconnect connectors, malfunction alarms, etc.).

Improper specifications. Even supposing that all necessary characteristics are specified, in an enforceable manner, there is the danger that the real intent is not accomplished. In this regard, the most common error is probably the "standard conditions" trap. Requiring that a particular parameter, or set of parameters, be measured under a standard set of conditions is fine, as long as we are certain of the parameters' variation as a function of those conditions. If a solenoid is to be driven with a 28-V 20-Hz square wave, little is obtained by requiring that it present an inductance of 50 mH or less to a 10-V 1000-Hz sine wave.

8.3.2 Part misapplications

If a perfect set of specifications for purchased materials, parts, and components exist, there is still the opportunity to err by ignoring specified characteristics.

Operation outside specified bounds. In some cases, the difference between misspecification and misapplication is somewhat foggy. For instance, in a case where we find ourselves overstressing a part, we could ask ourselves, "Why didn't we specify a higher rating of this parameter?" Well, probably because the specified rating is adequate for 99 percent of our applications of this part type—or the specified rating may represent the state of the art. In the category of design for operation outside specified boundaries we include such things as: (1) designing a circuit around the specified h_{fe} for a given transistor type, but at collector currents or temperatures either above or below the region for which the specified h_{fe} is valid; (2) exposing a natural rubber diaphram to ozone; or (3) exposing a hard but brittle mechanical member to high impact forces.

Failure to account for parasitics. Parasitic inductances, capacitances, mechanical inertia, etc., are often overlooked by the designer. They are easy to overlook because they are of no consequence in many applications.

Common parts application traps. Some of the more common electrical-part misapplications involve direct parallelling of active devices and "stacking" of voltage stand-off devices. These are forms of design for operation outside specified boundaries, but there is a subtle distinction to make. The fallacy here has to do with the part-to-part *variation* in specified parameters. If a pair of transistors had identical transconductance, then they *could* be directly parallelled to double the maximum collector current capacity. However, the specification undoubtedly allows a certain amount of part-to-part variation in transconductance. This leads to a very high probability of misapplication.

8.3.3 Design oversight

By having a good set of specifications and by assuring that each part in the design is assigned a proper function, we have gone a long way to ensure a reliable product. Nevertheless, in the interconnection of these parts into a performing system, there is still ample room for error.

Tolerance build-up. Part parameter variations can be additive, or even multiplicative, to the point where the intended equipment function is prevented. Part parameters vary with chance (purchase or fabrication tolerances), time, and environment (notable temperature), and this must be anticipated by the design.

"Race" conditions. The assumption that theoretically instantaneous occurrences actually happen instantly often leads to problems involving what is called a "race" condition. This is a condition wherein two or more events are supposed to occur at the same time, or in a particular sequence; otherwise the equipment will malfunction, possibly damaging itself, the operator, or the casual passerby. Automatic sequencing of power supply turn-on and turn-off is a common design technique for avoiding certain "race" conditions.

Stored energy. Stored energy often presents reliability (and sometimes safety) hazards, particularly in those instances where the storage of energy is a by-product of the desired function, rather than a conscious use of the energy storage phenomenon. The inductive spike which occurs when a relay is turned off is a good example.

"Primrose paths." A particularly beguiling path to design trouble is the "something for nothing" illusion offered by "tricky" solutions to design problems. For instance, one requires a very long time constant in a circuit application: A high-resistance capacitor charge path promises to provide this with a reasonably obtainable capacitance value. It is easy to overlook capacitor leakage currents, usually negligible, that can develop significant voltages in high impedance circuits—perhaps sufficient to prevent operation. Recognizing this, we might decide to use another trick—the Miller multiplier circuit—which makes a small capacitor in the base circuit of a transistor appear to be a large capacitor in the collector circuit. This large virtual capacitor, however, also has a large virtual leakage. The Darlington circuit offers fantastic current gains at a small price and in a small space, but some of this gain must be sacrificed lest we have thermal runaway.

We don't wish to detract from, or discourage, such clever solutions to design problems. Where their inescapable side effects are tolerable, they are very commendable. We do counsel caution to assure that the side effects are tolerable in each application.

"Bad luck." Even with good specifications, proper parts application, and careful circuit and equipment design, a finished product may suffer deficient reliability and maintainability. Certain parts or components may exhibit inordinately low wearout life. Hazards to personnel and equipment safety may appear. At first glance, these problems can only be attributed to just plain bad luck.

Investigation of such problems usually leads to the conclusion that there was indeed something that could have been done, in the nature of design/development program *planning and control,* to have prevented the problem. In short, "bad luck" is usually a sign of *undisciplined* design.

8.4 Practical Solutions

There are two major categories of practical design assurance techniques: design guidance and design review. Very simply stated, design guidance attempts to *prevent* design error, and design review attempts to *detect* and *correct* design error. This cannot be successful unless the guidance and review techniques are thoroughly grounded in a combination of experience and sound basic design assurance analysis of the design problem. The experience need not necessarily be that of the individual design assurance practitioner (although that is the ideal case). The necessary experience can be synthesized, by the conscientious student, from recorded assurance science history.

8.4.1 Basic design assurance analysis

"Basic design assurance analysis" is to practical design assurance engineering as pure mathematics is to science in general: "handmaiden to the sciences...may she never prove useful." In other words, no amount of exercising the basic analytical techniques alone will have any effect on the design. On the other hand, really effective design impact cannot be achieved *without* a sound analysis. The more common and useful analytical techniques are catalogued and described in the following chapter. Those referred to as "basic" include mathematical modeling, parameter allocations, and quantitative assessments (predictions and evaluations).

Mathematical modeling. In order to understand the implications of a given design approach, and particularly the implications of design alternatives, there is frequently the need for a mathematical model of some sort. The math model is an attempt to synthesize some set of product characteristics as functions of the *design*. It may be viewed as a receptacle for the results of the more detailed analyses (including allocations and assessments), in context with specific design approaches (system configurations, application environments, maintenance and logistics support philosophies, etc.). Math models come in several varieties, such as systems effectiveness models, life cycle cost models, logistics support (or sparing) models, reliability models, maintenance models, thermal models, stress models, etc. In addition, there are the less common models appropriate to peculiar problems—warranty cost models, for instance. Any or all of these may be pertinent to a particular design problem. All of them affect, and are affected by, product integrity parameters. None of them are "standard," to the extent that one specific set of equations is sufficient to describe the systems effectiveness of all systems. The math model must be developed specifically for the system of interest. Nevertheless, there are certain common principles to be applied in the model development. These principles are well documented in the trade literature.

In summary, a proper math model is the mechanism that allows realistic evaluation of the effect on some aspect of product integrity which results from the set of design characteristics. It can sometimes give surprising answers, defying intuition—answers which can be had by no other means short of disastrous, too-late, experience.

Quantitative integrity parameter allocations.   Allocation is the art of establishing useful "design-to" requirements for the constituent components of some larger system for which an overall requirement has been previously established. Allocations are necessary because it is otherwise impossible to design a component in such a manner that it and its brothers collectively meet the system design constraints.

Imagine a chief architect attempting to design a manufacturing facility with a total floor space of 20,000 ft^2. A junior architect is assigned to prepare layouts of the receiving and shipping areas, another to lay out the assembly area, another to prepare drawings of the purchasing and marketing office spaces, etc.—with no space constraints except the knowledge that the total space available is 20,000 ft^2. Most certainly, the resulting set of plans will describe a total floor space considerably in excess of the allowable total. Still worse, the relative space assigned to the various areas will probably have little relationship to the most effective utilization of total space for most efficient plant operation. Nevertheless, this is one way to get a starting point for space allocations. Then the chief architect's knowledge of other design considerations (e.g., anticipated material flow volume allows sharing of space by receiving and shipping; most marketing customer contact is to be "on the road" rather than "in-house," so that luxury and spaciousness of marketing offices is not of primary importance) might be super-imposed to scale down the space allocations for the various areas, in an equitable manner. In the more comprehensive case, a "floor-space math model" would be created which would consider the possibilities of having a multistory structure, various geometries, etc.

The objective, in formulating allocations, is to reconcile that which is required or desired with that which is achievable. All available ingenuity should be used to achieve allocations that describe reasonably attainable goals and result in meeting the overall design constraints. When this is not possible, relief should be sought from the design constraints. Trying to design according to allocations that are *difficult* to meet is a healthy thing. Kidding oneself along with allocations that are *impossible* is almost as bad as having no goals at all. Obviously, making the fine distinction between what is impossible and what is merely difficult requires a rare combination of experience and judgment. This partially explains why failure to develop and enforce appropriate product integrity parameter allocations is the single most serious and frequent failing of the design assurance function.

Product integrity assessments. "Assessment" is the broad term which is applied to those analytical techniques that seek to answer the question, "How 'good' is a particular design, with respect to a particular parameter (or set of parameters) of product integrity?" Product integrity assessments are performed at several levels of perspective and at several levels of detail. A systems effectiveness assessment, for instance, is from a very broad perspective. It is achieved by exercising the systems effectiveness math model, which has been "filled" with more narrow-perspective assessments of performance, reliability, and maintainability parameters. The level of detail has to do with the responsiveness of the model to the driving parameters, and the responsiveness of the performance parameter calculations (reliability predictions, etc.) to the independent parameters that affect them. The natural tendency is to proceed from the broad-perspective less-detailed assessment during the design concept phase—to the narrow-perspective more-detailed assessment during detailed design. This is how it *must* be (because of the evolution of design data) and—wonder of wonders—also the way it *should* be, because our "design to" criteria move from the overall requirements to the detailed allocations, as the design progresses.

Basically, assessment of integrity parameters is no different than assessment of performance parameters. Performance parameters, however, are usually deterministic, whereas most product integrity parameters are probabilistic. It is, therefore, ordinarily more difficult to establish credibility for predicted integrity parameters. Care must be taken to predetermine the assessment performance rules and the rules by which the results are interpreted. Above all, procedures for assessment must assure *consistent* results; i.e., methods must be scientific, repeatable, and consistent with cause-and-effect relationships.

8.4.2 Design guidelines

Design guidelines and ground rules are among the most important products of design assurance engineering. In many times and places, they are considered unimportant and are therefore ignored. This is partly because assurance specialists have habitually been either too preoccupied or too lazy to develop design guidelines and ground rules relevant to the design project at hand. Instead, reliance is placed on "motherhood" and "boilerplate."

On any new design project, relevant design guidelines and ground rules should be developed and disseminated. They should include answers to the following questions:

- What grades of parts should be bought for this product—commercial parts, MIL-standard parts, high-rel parts? What specific types are preferred? Who are the preferred vendors? What are the derating rules? What are the worst-case design tolerances?

- What design cautions are dictated by peculiar operating environments?

- What is the anticipated maintenance environment? What, in general, shall be designed to be the "replaceable unit" level for on-line maintenance? For shop maintenance? How will malfunctions be recognized and isolated?

- What are the reliability and maintainability requirements for components of the product? As the design progresses, how will judgment be made as to whether these requirements are being met?

From the very beginning of any design/development project, it is the person who gets there "fastest with the mostest" who is listened to. The design assurance engineer who can draw the quickest and soundest conclusions from the least detailed design information is in a position to have profound influence on the design. To make use of this advantageous position, remember the following rules:

1. Don't be timid in pointing out potential problems associated with various design aspects or approaches. Don't fear ridicule for pointing out "obvious" problems or asking "obvious" questions.

2. Convert all conclusions regarding product integrity to terms which are clear to nonspecialists.

3. Document all ideas concerning design ground rules and guide lines—
even if only in handwritten, dated, and signed notes. Send copies to
everyone who should be interested.

Parts selection and application. "A chain is only as strong as its weak-
est link." "The whole is equal to the sum of its parts." These old saws
(though they are often misapplied) illustrate the point that the proper
selection and application of component parts is essential to product
integrity. Part selection and application are not independent actions.
One must select parts that are suitable for the intended application.
One must apply the selected parts so that they are not asked to do something of which they are incapable. The parts selection and application
process can be viewed as consisting of three main sectors:

1. *Standardization.*   Confining design usage to the minimum variety of part types required to do the job; controlling the manner in
   which component parts requirements are specified. An important
   standardization tool is the "standard, or preferred parts list," which
   may apply to a particular project, product, or product line, or it
   may be company policy.

2. *Quality assurance.*   Assuring that parts and vendors are so selected
   and specified that parts quality control is possible during production. The major tools are the parts specifications themselves and
   such things as "qualified, approved, or standard supplier/vendor
   lists."

3. *Application ground rules.*   Assuring a proper and consistent set
   of constraints on the design application of component parts. Two
   important tools are stress derating and part parameter tolerance
   ground rules.

The relative importance of these three sectors is determined by a
number of product and project characteristics, as is discussed later.

The most important thing about the standard or preferred parts list
is that it should be enforceable. Occasional selection of parts not on
the list will be a necessity for most projects. The frequency of such occa-
sions, however, will be a direct function of the inadequacy of the list,
and inversely related to the enforceable procedural hurdles set in the
way of using nonstandard parts.

As for parts quality assurance, it would be nice if reliance could be
solely on specifying measurable part integrity parameters, and checking delivered parts for these parameters upon receipt. Unfortunately,
an instrument that can quickly measure the reliability or quality of
internal construction has yet to be invented. Reliance must, therefore,
be on a combination of (1) writing the specifications in such a way that

upon receipt it can be determined to the maximum possible extent whether the part complies and (2) careful selection and control of suppliers and vendors.

Application ground rules embrace prohibiting the use of certain part types in certain applications (e.g., natural rubber parts in the presence of petroleum distillates); design stress derating rules; and rules for designing to accommodate part parameter tolerances.

Stress derating is a technique whereby a component part is applied so that the stress values are less, by some margin, than the manufacturer's ratings for that part. By increasing this margin for mechanical, thermal, and electrical stresses, the possibility of degradation or catastrophic failure is lessened.

Designing to accommodate part parameter variations with chance (purchase tolerances), temperature (thermal coefficients), and time (effects of aging and wearout) is one way of recognizing that:

> the race is not to the swift, nor the battle to the strong,...but time and chance happeneth to all.
>
> Ecclesiastes 9:11

Be cautioned, however, that overzealous stress derating or tolerance allowances can very easily become "too much of a good thing." Designing to allow for the absolute worst combination of tolerances can result in a product which is not as reliable as it should be, under ordinary conditions, although it will still function under that one-chance-in-a-million set of conditions, should it ever occur. Predetermined stress derating and tolerance allowance ground rules should bear this in mind.

Hardware partitioning.   Too often, the first indication that anybody cares about the physical "building block" structuring of the product occurs at a formal design review, or worse yet, by means of a user complaint. Assuming that a reasonably complex product is being examined, we usually subdivide it into physical equipment groups, sets, assemblies, and subassemblies—in order to make the product handier for mass production (producibility), service and repair (maintainability), and logistic support (optimum life cycle cost). At the same time, the manner in which this subdivision is envisioned must account for the implications of product reliability, ability to withstand the environment, and aesthetic appeal. These interests are frequently at odds with each other, so that trade-off analysis is necessary in order to establish a proper physical packaging design philosophy (i.e., a set of ground rules) for *this product*.

One fundamental rule for packaging design, almost universally applicable (but which requires further interpretation for a particular product), is conformance to *functional packaging*. Functional packaging means making physical boundaries conform to functional boundaries. You create the physical boundary wherever you can draw a bound having the fewest and least ambiguous interface points around a discrete function.

Quantitative assurance budget. As the larger physical boundaries are established, corresponding budgets for product integrity parameters need to be developed. As physical subdivisions begin to take form in the design, the budgets are subdivided accordingly. These budgets are extensions of the quantitative integrity parameter allocations discussed earlier. Nonallocation, misallocation, or nonenforcement of allocations are the most frequent failings of the design assurance function. Part of the explanation for this has been that the derivation of allocations is a fine art, not yet developed to a full "science-hood." Another reason, appropriate for discussion here, is that allocations, even when artfully developed, are seldom translated into budgets. An allocation of reliability, for instance, that states a particular item must have a .999 probability of successfully operating for 24 h, has no particularly useful meaning to the item designer. Some of the fog lifts if the allocation is presented in terms of allowable hardware failure rate (or its inverse, mean-time-between-failures, if appropriate). Much more meaningful budgets, though, would be the answer to: "What does this mean in terms of maximum allowable total electronic parts count, active element count, maximum allowable temperature rise, minimum acceptable B_{10} life for bearings, etc.?" The raw allocations for other integrity parameters (MTTR, availability probabilities, system dependability, capability, and effectiveness indices) are likewise meaningless to the design engineer in terms of what must actually be done. The design assurance engineer has the responsibility of translating design integrity requirements into understandable design constraints, however difficult and tenuous this may be.

8.4.3 Design review

The next best method for preventing design errors is detecting and correcting them as early as possible, preferably before the design is released for manufacture. The design review process is the most effective known method for early detection of design errors. Design review is not merely a formal inquisition or a series of formal design review meetings. They are included as well as the many other activities that should be accomplished on a day-to-day basis as design progresses— the human factors engineer examining the preliminary console sketches, the reliability engineer comparing predictions for alternative designs, and the individual design engineer asking a coffee-buddy's opinion of a particular approach.

The design assurance engineer's attendance and participation in formal design review meetings is most effective when the necessary homework has been done. This homework consists of exercising the design examination techniques which are familiar to the design assurance engineer. These include parts application review/stress analysis, failure modes and effects analysis, worst-case analysis, and prediction/ allocation comparison.

Assurance specialists should resist doing *all* these things in excruciating detail, *all* the time. It simply isn't always cost-effective. On the other hand, techniques shouldn't consistently be avoided just because they are difficult or expensive. For example, many organizations never apply worst-case analysis unless it is contractually required by their customer.

Parts application review. Essentially, parts application review consists of ascertaining to what extent the parts selection and application guidelines are being followed. As it is developed, the parts list should be continuously compared with the preferred parts list; and justification for deviations should be critically examined. An independent (of the designer) calculation of electrical, mechanical, and environmental part stresses serves to verify conformance to parts application guidelines (or, in their absence, to sound engineering practice). Independent part stress analysis is automatic when worst-case analysis is a part of the development program—and of course it shouldn't be done twice. (It has happened.) In the usual development program, an independent part-by-part stress analysis for each and every part in the system is not recommended. It is usually almost as effective, and much more economical, to select a few high-population subassemblies from diverse areas of the equipment (e.g., power supplies, digital logic, gearbox, etc.) for stress analysis. If, in a particular area, stresses are high enough to arouse suspicion, one might then select a larger sample from that area for additional analysis.

The purpose of stress analysis is to see by examination of the design, if any parts are being stressed beyond, at, or near their limits of endurance—not to enable a more accurate reliability prediction. Even if the precise stresses on each part are known and recorded, it is wasteful to incorporate this information on a part-by-part basis into reliability predictions.

Failure modes and effects analysis*. One of the specific design analyses whose results are useful to reliability, maintainability, and system

---

*When failure criticality is considered, this analysis is referred to as a failure modes, effects, and criticality analysis (FMECA)—as discussed in Chapter 9.

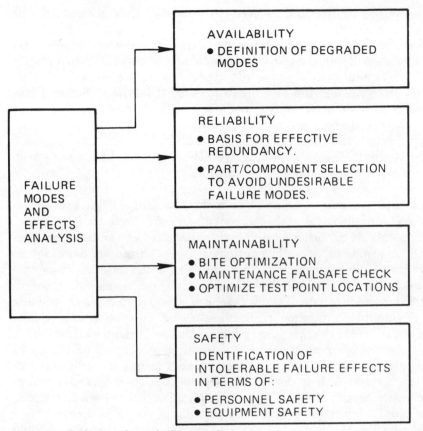

Figure 8.1   Failure modes and effects analysis.

safety studies (see Figure 8.1) is failure modes and effects analysis (FMEA). FMEA consists of inferring the failure mode of a higher level assembly (e.g., brake assembly) from a particular failure mode of one of its constituent parts (e.g., return spring). The process is continued until all significant failure modes of constituent parts are accounted for and the theoretical effects on higher levels of functional assembly are assessed until they are apparent at the desired levels.

FMEA is a powerful tool but is not appropriate or cost-effective for application to the development of all products. It is particularly appropriate to the development of products that are mission- or safety-critical, remotely operated, inherently difficult to maintain, or intended for high-volume production.

There are two "degrees of freedom" for FMEA: (1) the "level" of the analysis—i.e., whether one tracks failure effects from the piece-part to the system level, or from the subassembly to the major assembly level, etc., and (2) the "quantitativeness" of the analysis— i.e., whether one begins with the probability that a piece-part will fail in a given mode, or whether one guesses at the probability that a given assembly will fail in a particular manner and proceeds to the resulting probability that the system will fail in a particular mode, all the way down to a purely qualitative analysis. As you would expect the less detailed the "level" of the analysis, and the less "quantitative" it is, the easier and less expensive it is to perform. It follows that in the more expensive direction, the results become more exact and, therefore, more useful. This is true for smaller, less complex systems. For larger, more complex systems, a parts-level-up quantitative FMEA gets mind-boggling—even with computerized simulation programs.

Worst-case analysis. Perhaps the most powerful design evaluation method is detailed analysis of components to verify their ability to maintain acceptable performance under all anticipated extremes of the operating environment, as well as anticipated part parameter degradation with time and environment. This is called worst-case analysis and comes in several varieties, such as steady-state, transient, absolute worst case, and statistical worst case. Even though there is now a variety of computer programs available to assist in worst-case analysis, this is still one of the more expensive methods of design evaluation, as well as one of the more powerful. One hundred percent across-the-board application of worst-case analysis is not appropriate nor cost-effective for large systems. It is, however, very appropriate to smaller equipment, particularly if the equipment application is in any way critical. It is equally appropriate to critical portions of larger systems.

Having decided whether this kind of analysis is appropriate, we must then select the type. Dc analysis is usually sufficient for digital applications, particularly low-speed designs. Dc plus transient analysis may be required for power equipment. Ac or ac plus dc is usually appropriate to audio and rf circuitry. Whether the absolute or statistical worst-case approach is taken with any or all of these is another question. Generally, the absolute worst case for the more critical failure consequence situations, and the statistical worst case for the less critical but high-production-volume situations, are recommended.

Prediction/allocation comparison. Periodic comparison of reliability, maintainability, and availability predictions with corresponding allocations is a reasonably straightforward operation. However, when the comparison is unfavorable (i.e., the reliability, maintainability, and availability predictions are not as good as the allocated), there is a general tendency to immediately rearrange the allocations. An allocation should be treated as firm, unless there has been some postallocation development which negates the boundary conditions or logic of the allocation. Also, care should be taken to compare the most meaningful parameters. For instance, although the most meaningful design-to-reliability allocations are probably MTBF and associated parts count budgets, the most meaningful examination of how far the design is off might be in terms of required vs. predicted mission reliability.

All of the foregoing design guidance and review techniques, in conjunction with disciplined design activity, should culminate in successful formal design reviews. Yet in order to assure this success, more must be known than just these techniques and their use.

8.5 Tailored Design Assurance Programs

In spite of everything, deficiencies will creep into equipment designs. Most of these deficiencies are of the sort that crop up time after time, and that the standard design assurance techniques are supposed to prevent or correct. The standard techniques can, in fact, be very effective in minimizing these deficiencies—if properly applied. Effective design assurance comes from the appropriate set of these techniques applied to a specific product.

8.5.1 Characteristics of effective design assurance

Perhaps the single most important element of effectiveness for all of the design assurance techniques is *timeliness*. Even the simplest, riskiest, and most obvious conclusions and recommendations are better than *no* information at points where design decisions must be made. Even the most profound and confident conclusions are of little use after manufacturing resources have been committed to a finished design. Hastily drawn, erroneous conclusions can cause a lot of grief. However experience dictates that grief is caused much more frequently by too little, too late, in the way of design direction and evaluation.

Just as important as timeliness, but probably less often ignored, is *pertinence*. The technique and the manner of its application must be appropriate to the design problem, as well as to the current stage of solution of that design problem. For a remotely operated, periodically shut-down system, for instance, hardware packaging should probably be directed more toward minimization of interconnections for reliability rather than modularity for maintainability.

Finally, there is the need for the "design-active" end of design assurance technique applications (published guidelines, corrective action recommendations, etc.) to be *provocative*—to get the attention of those who are in a position to make it happen. The design assurance engineer and organization must build a reputation for sagacity and veracity and be willing to "lay it on the line." Design assurance personnel must be willing to take their case to the highest accessible authority.

8.5.2 Technique selection and application

The first step in selecting a set of design guidance/evaluation techniques to be applied on a given program is to anticipate what kind of design deficiencies are most likely to occur and which are most likely to hurt the most. Having decided this, one could use some arrangement, such as the deficiencies vs. techniques matrix of Table 8.1, to select the most appropriate techniques. This table identifies the kinds of design deficiencies that each technique is most likely to prevent or detect. However, because it is difficult to anticipate problems by class, one must really back up for a practical first step. The relative likelihood of occurrence of the various design errors, and the seriousness of their consequences, is related to a number of program and product characteristics. Some of these program/product characteristics also have a bearing on technique selection and application—other than through the generation of deficiencies. Intended high-production volume, for instance, has the effect of making in-depth application of all the design guidance/evaluation techniques more appropriate. Not only does this provide a larger sales base over which to amortize design-related activities, it increases the payback for any deficiencies prevented or corrected through the use of these techniques.

Three product/program characteristics (product application criticality, product complexity, and production volume) have been chosen to exemplify design assurance technique selection/application criteria (see Table 8.2). In this table, the selection and application of techniques have been related directly to these product/program characteristics. Other product/program characteristics such as maintenance philosophy and design schedule might have been included—but the table complexity and physical dimensions increase exponentially with the number of characteristics used as criteria. Table 8.2, therefore, should be viewed as illustrating the idea that there is a predictable relationship between product/program characteristics and the proper selection and application of R/M techniques; it should not be seen as a complete guide to the actual selection and application. The development of a more comprehensive matrix is left to the interested student.[*]

---

[*]A very similar approach to design assurance technique selection (a quantitative method, but applied specifically to reliability assurance) appears in *Proceedings of the 1974 Annual Reliability and Maintainability Symposium* (IEEE Catalog Number 74CH0820–1RQC), by J. E. Bridgers, Jr.

TABLE 8.1 Common Design Deficiencies vs. Preventative/Corrective Techniques

Category	Deficiency	DESIGN GUIDELINES	PARTS SEL. & APPLICATION	PREFERRED PARTS LISTS	STRESS DERATING RULES	END-OF-LIFE TOLERANCES	HARDWARE PARTITIONING	R/M BUDGETS	DESIGN REVIEW	PARTS APPLICATION REVIEW	FAILURE MODES & EFFECTS ANAL.	WORST CASE ANALYSIS	PREDICT./ALLOC. COMPARISON
"BAD LUCK"	SAFETY HAZARDS						X				X	X	
"BAD LUCK"	POOR MAINTAINABILITY						X	X			X		X
"BAD LUCK"	SHORT LIFE			X	X								
"BAD LUCK"	LOW RELIABILITY			X	X	X		X			X	X	X
DESIGN OVERSIGHT	PRIMROSE PATHS											X	
DESIGN OVERSIGHT	STORED ENERGY											X	
DESIGN OVERSIGHT	"RACE" CONDITIONS											X	
DESIGN OVERSIGHT	TOL. BUILD-UP				X	X						X	
MIS-APPLICATION	TRAPS									X		X	
MIS-APPLICATION	PARASITICS/TRANS				X					X		X	
MIS-APPLICATION	OUTSIDE SPEC				X					X		X	
MIS-SPECIFICATION	UNENFORCEABLE	X						X					
MIS-SPECIFICATION	IMPROPER	X										X	
MIS-SPECIFICATION	INCOMPLETE	X						X				X	

Technique groupings (bottom brackets):

- END-OF-LIFE TOLERANCES → BASIC R/M ANALYSIS
- HARDWARE PARTITIONING / R/M BUDGETS → MATH MODELING, ALLOCATIONS
- PARTS APPLICATION REVIEW → PREDICTIONS

TABLE 8.2 Design Assurance Technique Selection and Application Matrix

KEY TO APPLICATION DEPTH:
Quick and dirty → Intensive & careful

Notes spanning the full width of the matrix:

- **PREDICTIONS / BUDGETS (row note):** Down to the series function level; given in terms of physical design factors such as parts count, transistor count, fraction of faults detectable, etc.

Design Assurance Technique	NON-CRITICAL PRODUCT MISSION, REASONABLY BENIGN APPLICATION ENVIRONMENT — Low Complexity Product (few/simple functions) — LOW	— Low Complexity — HIGH	— High Complexity, Multi-Function, Multi-Mode — LOW	— High Complexity — HIGH	PRODUCT FAILURE HAS SERIOUS CONSEQUENCES, APPLICATION ENVIRONMENT IS SEVERE — Low Complexity Product (few/simple functions) — LOW	— Low Complexity — HIGH	— High Complexity, Multi-Function, Multi-Mode — LOW	— High Complexity — HIGH
QUANTITATIVE ANALYSIS — MATH MODELING	Simple, conservative model.		Account for all redundancies, "give away" alternate mode advantages, if it saves time and effort.		Account for all modes, tie in with FMEA where possible. Applicable environments.		Account for all functions, modes & environments tie in with FMEA if possible. Account for effects of repair on Rel., Avail.	
ALLOCATIONS	Sufficient to establish meaningful reliability budgets.		Sufficient to establish meaningful R/M budgets. Establish bounds for fault detection & isolation.		Firm allocations for all R/M parameters under all environmental & application conditions.		Derive in consonance with math model. Bound fault detection & isolation.	
PREDICTIONS	Average part stresses by type. Average internal temperature. Applicable environments.				Part stresses by type sub-group, by equipment area. Temperature by equipment area. Applicable environments.		Average part stresses by type and by equipment area. Temperature by equipment area. All applicable environments.	
DESIGN GUIDELINES — PARTS SELECTION & APPLICATION (PRIORITIES)	1. Q.A. 2. Application 3. Standardization		1. Q.A. 2. Application 3. Standardization	1. Standardization 2. Q.A. 3. Application	1. Application 2. Standardization 3. Q.A.		1. Application 2. Standardization 3. Q.A.	1. Application 2. Standardization 3. Q.A.
HARDWARE PARTITIONING	For economical build & sparing.		To allow non-interfering repair, enhance economical sparing, non-ambiguity of fault isolation.		To enhance fault isolation, remove/replace rapidity.		To allow non-interfering repair, non-ambiguity of fault isolation, enhance remove/replace rapidity.	
BUDGETS	Down to the series function level; given in terms of physical design factors such as parts count, transistor count, fraction of faults detectable, etc.							
DESIGN REVIEW — PARTS APPLICATION REVIEW	Spot-check of part stresses and applications.		Spot-check of part stresses and applications, by equipment function categories.		Comprehensive stress analysis and verification of adherence to selection & application rules.		Reasonably extensive spot-check of stresses; comprehensive verification of selection and application.	
FAILURE MODES & EFFECTS ANALYSIS	NONE	Comprehensive, parts-level-up, qualitative FMEA.	Selective, qualitative, gross level FMEA.	Selective, qualitative, mid-level gross FMEA.	Comprehensive estimate-quantitatives, FMEA.	Comprehensive quantitative, parts-level-up FMEA.	Comprehensive, qualitative, mid-level FMEA.	
WORST CASE ANALYSIS	NONE	Very selective, include most critical assys. only; Statistical WCA.	NONE	NONE	Selective, including all critical areas of system; Absolute WCA.	Complete and comprehensive, Absolute WCA.	Very selective, include most critical assys. only; Absolute WCA.	Selective, including all critical areas of system; Absolute WCA.
PREDICTION/ALLOCATION COMPARISON	Straightforward comparison, allocations firm.		Compare at all levels, i.e., hardware and functional. Allocations somewhat flexible.		Straightforward comparison, allocations firm.		Compare at all levels, i.e., hardware and functional. Allocations somewhat flexible.	

Table 8.2 indicates the appropriate depth of analysis and the appropriate technique. The appropriate comprehensiveness (with respect to the number of techniques to be applied, as well as the fraction of the system to which each technique should be applied) is largely a function of product application criticality and production volume. The depth of analysis, necessary in any case, is a function of the degree of confidence one requires in the results. Therefore, experience or judgment can be used to supplement or supplant analytical rigor, if need be. Also, this characteristic makes it generally appropriate to begin an analysis by making it a rather gross but comprehensive effort. Suspicious areas can then be examined more deeply, to develop the desired confidence that there is, or is not, a problem. Since the confidence required of analytical results is related to the consequences of being wrong, required depth of analysis is obviously directly related to product application criticality.

In summary, the successful design assurance engineer must (1) *understand* the design problem at hand, (2) know what kind of design deficiencies to *anticipate*, (3) be able to *translate* product integrity requirements into physical design and support package terms, and (4) be adept at effectively *communicating* these requirements.

8.6   Case Histories

8.6.1   The case of the up-tight solenoid

Once upon a time, there was a solenoid whose mission was very important. The solenoid's basic job was to mechanically store (via a ratchet mechanism) information related to the number of 28-V square pulses it received during a particular sequence of events. As it turned out, it counted any number of pulses as just one. It simply pulled in and never relaxed until the pulse train ceased.

The solenoid driver circuit had been designed around the solenoid specification, a part of which required that the solenoid coil inductance not exceed 50 mH. The driver circuit designer accounted for this inductance (including tolerance margin) in assuring that the coil current would decay well below the solenoid holding current (even with the necessary arc-suppression arrangement to protect the drive transistor) during the off times of the 20-Hz pulse train.

The problem was the fine print. The solenoid specification required the coil inductance to be not greater than 50 mH *when measured with a 10- V 1000-Hz sine wave.* Subsequent tests indicated that the effective energy storage inductance under operating conditions was in excess of 300 mH. The solenoid would not react to off times less than the half cycle of a square wave of 12 Hz or less.

A case of design specification not matching the specific design application—part misspecification.

8.6.2   The case of the garroted klystron

The klystron was never actually hanged but was sentenced to hang. It was a very large and heavy device so that the maintenance planners had provided for special handling equipment, including an overhead winch and sling for lifting by two lifting eyes mounted on the klystron body. The position of the lifting eyes was carefully calculated, taking into account the center of gravity, so that the unit would hang precisely vertical. Of course this was important so that the maintenance people could get the device in and out of its socket.

Maintenance engineering analysis of the drawings revealed that accessories mounted on the tube would interfere with the hoisting sling in such a way that the tube could not hang vertically. Relocation of the lifting eyes was required.

A case of design oversight and assurance analysis "catching" the oversight.

8.6.3   The case of the exploding relay

The leader of a design team (which was engaged in final debugging of the prototype of a large test console which the team had designed) was strolling through the test area. He was, perhaps, congratulating himself that the end of this project was in sight. His reverie was shattered by a loud noise and the breath of passing shrapnel.

After the excitement had abated a bit, it was discovered that a small relay, one of a bank of 11, had blown right off a panel. The bank of relays were used to set up various test situations. A corresponding bank of pushbutton switches activated the relays. If a given pushbutton were pushed, it connected voltage to the corresponding relay, and that relay's contacts made the required connections for a particular set-up at the console's test connector. The pushbuttons were mechanically interlocked so that only one switch could be closed at a given time. Therefore, only one relay could be closed at a given time—right? Wrong! The relay dropout delay is greater than its pull-in delay. Therefore, starting with one relay closed and actuating another—the relay pulling in wins the race to change states.

A clear case of one type of race condition—often overlooked during design.

8.6.4 The case of the foreign service power supply

In each of a number of phased array radar systems installed in a number of foreign locations, there were tens of thousands of power supplies, one for each array element. The systems were delivered and installed in their locations about 6 months apart, operating 24 h/day, every day. The design and manufacture of the power supplies was subcontracted by the radar system developer.

Knowing that power supplies are often the least reliable element in any system, the system developer incorporated extensive assurance requirements into the subcontract statement of work. They included: (1) an extensive parts control program, (2) extensive functional and environmental testing requirements, (3) a specific design requirement that no single part failure could cause a "cascading" of other part failures within the power supply, and (4) a requirement that the power supply be completely sealed and potted—a throwaway device.

Almost precisely 1 year from turning on the power for the first installed system, a few power supply failures began to occur. These first few turned into a flood of failures. Analysis of the first few written failure reports by the system developer's assurance specialists indicated a high probability of an open capacitor causing output transistor failures. A high-priority request was made to obtain two failed power supplies from the foreign site for physical failure analysis confirmation. The paper analysis was confirmed. The specific cause of the capacitor failure was found to be electrochemical corrosion of the anode internal lead, which eventually led to an open circuit and consequent transistor failures. The corrosion resulted from chemical contamination which occurred during the capacitor manufacture.

The end result of all this was a complete field retrofit of all power supplies in all foreign locations with repaired power supplies, now—of necessity—made repairable. An extremely costly experience for the customer, system developer, power supply subcontractor, and capacitor manufacturer (they all shared in the total cost).

This is a very clear example of insufficient design analysis and confirmation of meeting the assurance requirements of the system. A design review program and failure modes and effects analysis would have had a very high probability of eliminating the problem while still in the design phase—a *much* cheaper solution. The capacitor might still have failed, but the failure would not have resulted in complete power supply failure.

9

Analytics

I think and think for months and years.
Ninety-nine times, the conclusion is false.
The hundredth time I am right.
ALBERT EINSTEIN

9.1 Concerns

Over the years, a number of analytical techniques have been found to be particularly useful in the pursuit of product integrity. They are the tools used to draw conclusions and make possible sound decisions relative to the following product assurance analytical concerns.

Requirements. What must we have from our product?
Capability. What can we expect from our product?
Conformance. How has our product done?
Improvement. What can we do to enhance our product?
Risk. Are we headed for trouble with our product?

From an analytical viewpoint there are both differences and commonality between these concerns and the techniques employed to address them. These concerns are reviewed as they focus primarily on the hardware segment of a product. Software analytical tools are discussed in Chapter 13, Software Quality Assurance.

9.1.1 Requirements analysis

You have to know what you want from a product before you can design and build it. The customer's requirements, whether perceived or specified, must be allocated or translated into a set of clear requirements suitable for your own use. In the absence of clear-cut customer requirements, this analytical process becomes even more important.

9.1.2 Capability analysis*

Once you know the product requirements, then you have to have some way of knowing where you're headed. In other words, you have to be able to predict the capability of your product relative to the established requirements. The process involves taking a product design, at its various stages of synthesis, and projecting what the product will be capable of doing. These results are then compared to the requirements to determine areas of existing or potential noncompliance.

9.1.3 Conformance analysis

After the product design is finished, it is translated into hardware. Now comes another opportunity to apply an analytical process, this time to assess the conformance of the product to its requirements. This conformance analysis can and should be performed both prior to and after delivery of the product to the customer. As in the case of capability analysis, analysis of product conformance is more than a one-shot proposition. For example, conformance of the product could be evaluated during the prototype stage. As the product hardware becomes finalized, it is subjected to further analysis to ensure that it conforms to its requirements. Finally, after the customer has the product, it can be evaluated for conformance to the requirements during actual use.

9.1.4 Improvement analysis

What do you do if your product isn't going to, or doesn't conform to the requirements? You improve it so it will meet the requirements. Toward this end, then, analyses are conducted to determine the best approach for enhancing the product's integrity and meeting the established requirements.

9.1.5 Risk analysis

Overlying the above analytical concerns is perhaps the most important analysis of all—risk analysis. Risk analysis can be directed toward either the existing or the impending disaster that the product could

---

*"Capability analysis," as used here, should not be confused with the analysis of process capability as it concerns machinery, materials, instruments, etc.

cause. In its narrowest form, risk could be related to the product's inability to meet its product integrity requirements. In a broader sense, risk could be tied to any problems that the product could create; e.g., even though safety requirements may not be specifically established for a product, the producer may still be liable for injury, death, or damage attributed to a product. Hence, risk analysis needs to begin early in a product's life and be continued as necessary through its later stages. And obviously, it is not limited just to the hardware product, but also extends to include software and the total product.

The remainder of this chapter is devoted to describing some of the analytical tools used by the product assurance professional. Because of the tremendous scope of a topic such as analytics, the treatment here is in smorgasbord fashion. The focus is more on the quantitative tools—those that are in widespread use by companies large and small, whether involved in defense or commercial products. Brief summaries are also provided for some of the other quantitative and qualitative tools not highlighted. And as indicated at the outset, the emphasis will be on techniques applied to the hardware product.

9.2 Foundation

Before describing the various analytical tools, we need to examine two fundamental topics—probability and statistics. Since these topics in themselves are generally the subjects of complete textbooks, no attempt is made to present a detailed discourse. Instead, the goal here is to provide an understanding of the basic principles to aid the reader later in the chapter as each analytical tool is described.

9.2.1 Probability

Events and trials.   Probability is usually associated with events; i.e., the proportion or percentage of time an event would be expected to happen in a large number of attempts or trials. This relationship leads to the so-called relative-frequency definition of probability—if an event occurs in s out of n trials and doesn't happen in the remaining $(n - s)$ attempts, the probability of occurrence is the limit of s to n, as n goes to infinity. Hence, in a well-designed experiment, with n being large, a fair approximation of the probability can be derived by the ratio s/n.

Another definition of probability is related to a count of the number of possible results of a trial. The simplest way to illustrate this definition is through the use of a single die. A die has six sides, so a single throw has six possible outcomes with the probability of any one number occurring is $\frac{1}{6}$. Generally speaking, if an event can occur in m ways and fail to occur in n ways, the probability of occurrence is $m/(m + n)$, given that the following conditions hold regarding the outcomes.

1. They are equally likely—that is, there is no bias which provides an advantage of one outcome versus another in a trial (the dice aren't loaded!)
2. They are exhaustive—that is, all possible outcomes of an event are included.
3. They are mutually exclusive—that is, when one thing is known to happen, the other is known not to happen; e.g., if the event being considered is the drawing of a single card from a deck and obtaining the king of clubs, there is only one mutually exclusive way of doing it; and if the king of clubs is in fact drawn, it can be no other card.

When speaking of trials, another factor to be considered is the independence of the trials. For example, each roll of the dice in a game of craps is independent, since the outcome of one trial has no effect on the outcome of the next trial. By contrast, if we fail to replace a card drawn from a deck prior to the next draw, we lose independence. For example, the probability of drawing a king of clubs initially is $\frac{1}{52}$; if we fail to draw the king of clubs on the first draw, and fail to replace the card we just drew, the probability of drawing the king of clubs on the second draw improves to $\frac{1}{51}$.

Probabilistic ideas.   Other ideas to be considered when discussing this topic are conditional or unconditional probabilities and discrete or continuous probabilities. The probability of an event is said to be conditional if we know something about the occurrence of some related event. This is particularly important in the case of dependent events (e.g., switching in a redundant element only upon failure of the primary element), where knowledge that one event has occurred does affect the probability of the other event. One the other hand, with lack of information concerning a previous trial, we speak in terms of unconditional probability. Discrete probability is associated with go/no-go types of events such as drawing and inspecting samples of a product to determine those which are good or bad. By contrast, continuous probability relates to the case in which numerical values can be assigned to the events—these values being any number which falls between two limits. An example of the continuous probability case is the probability of a television set operating properly for some interval of time.

Some useful probability rules. Two rules are particularly useful to the reader working with probabilities—the additive rule and the multiplicative rule.

Additive rule. If we have two mutually exclusive events, A and B, using the additive rule, the probability of either A or B occurring in one trial is the probability of A occurring plus the probability of B occurring. As an illustration, consider the roll of dice again. If A is the probability of throwing a 2 and B is the probability of throwing a 6 on a single throw of two dice, then the probability of throwing either a 2 or a 6 is given by

$$P(A) + P(B) = \tfrac{1}{36} + \tfrac{5}{36} = \tfrac{1}{6}$$

For the case where A and B are not mutually exclusive, the probability of either A or B is equal to

$$P(A) + P(B) - P(AB)$$

where $P(AB)$ is the probability of both A and B.

Multiplicative rule. If we have two independent events, A and B, then the probability of both A and B occurring is the probability of A occurring times the probability of B occurring. (Note: This rule can be applied to the case of more than two events, by extending the product to encompass all events.) For example, consider the deck of cards—if A is the probability of drawing a king of clubs and B is the probability of drawing an ace of spades on two successive draws (with card replacement after the first draw), then the probability of drawing both a king of clubs and ace of spades is given by

$$P(A)\,P(B) = (\tfrac{1}{52})(\tfrac{1}{52}) = \tfrac{1}{2704}$$

Combinations and permutations. Continuing further in the fundamentals, there are probability analyses which are concerned with the *number of favorable* ways in which an event can occur out of the *number of total* ways in which that event can occur. To attack this type of problem, one first needs to know the total number of ways or combinations (C) that n things taken m at a time can occur. Mathematically this can be expressed as

$$C(n, m) = \frac{n!}{m!(n - m)!}$$

By way of a very simple example, consider three black boxes operating in parallel with any two being required to successfully fulfill mission requirements. The number of ways in which the mission could be successfully accomplished is

$$C(3,2) = \frac{3!}{2!(3-2)!} = \frac{3 \cdot 2 \cdot 1}{(2 \cdot 1)(1)} = 3$$

Another consideration in probability analysis is the number of ways in which a group of m things selected from a set of n can be arranged. This is referred to as a permutation and is denoted by

$$P^n_m = \frac{n!}{(n-m)!}$$

In some cases, we only have two possible outcomes for each trial. For example, if a product were inspected, it could be considered either good or bad. The probability determination for these types of events is through the use of a binomial expansion relationship:

$$(p+q)^n$$

where p = probability of being "good"
q = probability of being "bad"
n = number of trials

9.2.2 Statistics

Random variables. In product assurance work, we deal with both discrete and continuous random variables. Discrete random variables are the go/no-go types of variables such as the number of rejected units in a lot. On the other hand, continuous random variables are variables such as time-to-repair observations.

Density functions. Density functions are applied in determining the probability that a certain value of a random variable will occur. This probability is expressed as a function of that variable. For a discrete random variable, the density function is obtained through a summation process. On the other hand, the density function for a continuous random variable is derived through an integration process. As an example, Figure 9.1 presents the equation and curve for the exponential density function—a commonly applied function in reliability analysis. Similar equations and curves exist for other types of continuous and discrete functions.

Distribution functions. Distribution functions are used to determine the probability that in a random trial, the random variable is or is not greater than a given value of that variable. A distribution function can be specifically tailored to the data being analyzed. Some of the common

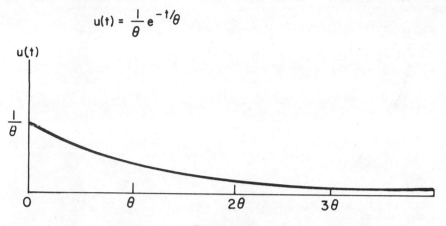

Figure 9.1   Exponential distribution density function.

functions, and the type of random variables they are associated with, used in product assurance work include the following:

Functions	Random variable type
Binomial	Discrete
Poisson	Discrete
Gaussian (Normal)	Continuous
Exponential	Continuous
Weibull	Continuous

As a comparison with the density function, Figure 9.2 shows the appropriate equation and curve for the reliability function associated with the exponential distribution. The latter distribution function is used to determine the probability of success over a given time t designated as $R(t)$. Similarly, for discrete random variables, an applicable distribution function could be selected and the associated equation employed to determine the probability of x successful trials, designated as $R(x)$.

Sampling theory.   In product assurance work, sampling becomes a very important approach toward ensuring product integrity. Sampling is employed in cases where the test of a product or its elements is destructive in nature (e.g., flash bulbs); in cases where large quantities to be inspected preclude the inspection of all items because of cost considerations; in cases where 100 percent inspection causes inspector fatigue which actually results in lower product assurance; and in cases where

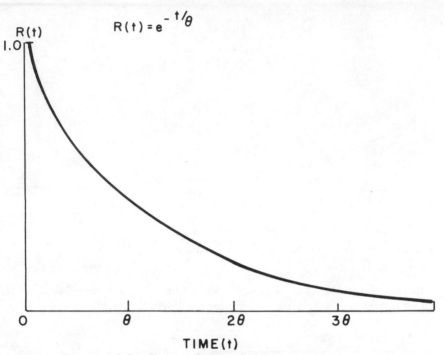

$$R(t) = e^{-t/\theta}$$

Figure 9.2 Exponential distribution reliability function.

it is impractical to assess the capabilities of a product under simulated
or actual operational use (e.g., it would not be economically feasible to
measure the average life of TV sets used in the home). Instead, for
reasons such as destructive testing, inspector fatigue, or economics, it
may not be feasible to inspect or assess the capabilities of every item
produced. Thus, we must resort to sampling, as opposed to studying
and assessing the entire population from which the sample is drawn.
Fortunately, there are some appropriate statistical measures that can
be calculated to lead to a conclusion about the population.

One useful statistic that can be calculated from a set of sample
measurements is the point estimate—a single value used as the best
estimate of the population value. In an attempt to lend some statistical confidence to inferences made from samples, the single point estimate is replaced or supplemented by an estimate of the interval
between which the population parameter lies. As a simple example,
consider the case in which a sample of 100 firecrackers is drawn from
a lot of thousands. Obviously a sample is in order, since the destructive nature of a firecracker test would leave none to sell. Upon firing
off the 100 units in the sample, it is found that five fail to go off. What
can we say about the population from which the sample was drawn?

Point estimate. The point estimate for the sample of firecrackers is simply

$$p = f/n = 95/100 = .95$$

That is, our best guess of the population is that the "true" fraction of good firecrackers in the lot is .95 (conversely, .05, or five percent are bad).

Interval estimates. Carrying this example further, we could also determine intervals within which the true population parameter is located. These intervals, or confidence intervals as they are called, come as one-sided or two-sided limits and take the size of the sample into consideration. Any statement that we make about the interval limits is related to the degree of confidence that the parameter is what we say it is.

Thus, we could say that we are X percent sure that at least Y percent of the firecrackers in the lot are good based on the sample results. A two-sided interval would yield a range within which the parameter is located. For this case, we could say that we are X percent sure that between Y and Z percent of the firecrackers are good.

Hypothesis testing. As discussed in Chapter 11, a very important concept in quality control and acceptance testing is that of hypothesis testing. Through this process, a hypothesis is stated concerning population parameters, data collected and analyzed, and a conclusion made as to whether the hypothesis is rejected or accepted. Thus, one might want to know if a process change made on integrated circuits substantially improved product yield. The hypothesis to be tested would be that there is no difference in the yields before or after the process change. This no-difference hypothesis is called the null hypothesis. We then set about gathering and analyzing data to arrive at a conclusion. It should be noted that sometimes an incorrect conclusion can be reached. That is, there can be type I errors—rejecting the hypothesis as false when it is in fact true; conversely, there can be type II errors—accepting the hypothesis as true when it is in fact false.

The next section of this chapter provides an opportunity to see how some of these fundamentals contribute to an understanding of product assurance analytical tools.

9.3   Tools

This section of the chapter presents descriptions of the more widely employed product assurance analytical tools. You won't be an expert after you've read this material. Hopefully you will have an understanding of how these kinds of tools contribute toward ensuring product integrity.

Included in this section will be discussions on (a) prediction, (b) allocation, (c) assessment, (d) failure modes, effects, and criticality analysis, (e) hazard analyses, (f) fault tree analysis, (g) control charts, and (h) experimental design.

9.3.1 Prediction

A very important tool to the product assurance practitioner, particularly one who is involved in reliability or maintainability engineering, is prediction. The prediction process is useful in that the results can be used (a) to tell us where we are relative to a requirement such as MTBF, MTTR, or inherent availability, (b) to identify potentially weak areas which require improvement, and (c) to evaluate risk areas. As used here, the word "prediction" will apply to those estimating techniques applied during the design phase to determine reliability and maintainability parameters (e.g., MTBF) of a product. Those estimating techniques that are used to assess the capabilities of a product during the test/operational phase (after hardware is built) are discussed in the section titled "Assessment."

The procedures that one applies to reliability/maintainability predictions are extremely varied. For example, in the case of reliability prediction, one can employ techniques that (a) relate the product reliability being predicted to one which is similar and for which historical data is available, (b) consider the expected failure contributions of the parts of the product and combine these contributions mathematically, or (c) take elemental building blocks of the product and relate average failure rates to these blocks. Similarly, for maintainability prediction, one could go through a maintenance-task time analysis in which each step in the product maintenance procedure (e.g., removal or replacement of a failed element) is evaluated and a time estimate placed against it. Whatever the technique selected, the following general steps have to be followed.

Reliability. *Define the product in as much detail as possible.* Unless every element that can possibly contribute to product failure is accounted for in one way or another, a true picture of its reliability will not be obtained. Some typical information would be (a) functional block diagrams, (b) parts lists, (c) schematics, (d) product operating descriptions, and (e) logic diagrams.

Establish a clear understanding of how the product functions. Is the product functionally interconnected or related—like the old-fashioned Christmas tree light strings, which failed completely when a single bulb failed; or is the product highly redundant—like a gigantic telephone system so that even if one line goes down, many paths are available for a telephone call to go through? The aim here is to determine what constitutes failure and success. The classical way that the product functions are portrayed from a reliability viewpoint is through the use of a reliability block or success diagram. This diagram shows either (a) the combination and sequence of events that must occur successfully in order for the product to do its job or (b) the relationship of the various product elements to each other. Figure 9.3 shows some very simple reliability block diagrams.

Develop a mathematical model which translates the product functions into terms that can be quantified to estimate its reliability. The model will generally be as simple or complex as the functions the product will be required to perform. Additionally, the model will become further complicated by the existence of multiple paths (redundancy) to reliable product operation.

The way the model will be structured will depend on the way the reliability requirements are specified. For example, the model for products such as electronic equipment, which are time-oriented in their operation, will be developed in probabilistic terms as a function of time or in terms of failure rates that are easily translatable to MTBF. On the other hand, for mechanical or "one-shot" devices (e.g., pyrotechnics), the model may be developed in probabilistic terms as a function of trials or cycles of operation.

It is here in the development of such models that a knowledge and understanding of probability and statistics becomes extremely important. If we take another look at the two simplest block/success diagrams shown in Figure 9.3 (a series string of two elements and a parallel arrangement of two elements), we can illustrate this point.

Series.   From the probability rules described earlier in this chapter, the model for the series diagram would be developed using the *multiplicative rule*. Therefore, the simple probability model would be

$$P_s = P(A)P(B)$$

where P_s is the probability that both elements A and B, hence the product, will function successfully.

However, the proper distribution function must be applied, if we are to be able to quantify this model later. Thus, assuming an exponential distribution function (see Figure 9.2), the model would be

$$P_s = P(A)P(B) = \exp\left(-t_A/\theta_A\right)\exp\left(-t_B/\theta_B\right)$$

SERIES: BOTH EVENTS/EQUIPMENTS A AND B MUST OCCUR/OPERATE SUCCESSFULLY

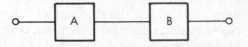

PARALLEL: EITHER EVENTS/EQUIPMENTS A OR B MUST OCCUR/OPERATE SUCCESSFULLY

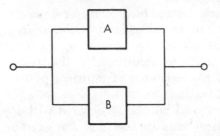

SERIES-PARALLEL: EITHER EVENTS/EQUIPMENTS A AND B OR C AND D MUST OCCUR/OPERATE SUCCESSFULLY

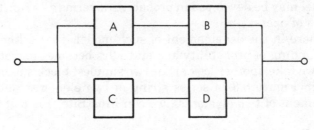

PARALLEL-SERIES: BOTH EVENTS/EQUIPMENT A OR C AND B OR D MUST OCCUR/OPERATE SUCCESSFULLY

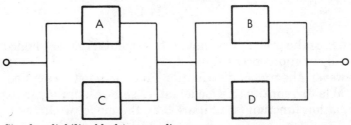

Figure 9.3 Simple reliability block/success diagrams.

where θ_A = the MTBF of element A
  t_A = the required operating time of element A
  θ_B = the MTBF of element B
  t_B = the required operating time of element B

Carrying this simple modeling example further, if we wanted to express the series elements in terms of their failure rates (i.e., expected failures per unit time), then the model would be

$$\lambda_s = \lambda_A + \lambda_B$$

where λ_s is the total failure rate of our two-element product, which would translate, in MTBF terms, to

$$\theta_s = 1/\lambda_s$$

where θ_s is the MTBF of the overall product.

Parallel. Again drawing from our earlier probability rules, we would use a model which reflected the *additive rule*. Therefore, the probability model (P_p) for the parallel arrangement would be

$$P_p = P(A) + P(B) - P(A)P(B)$$

For quantification purposes, again assuming an exponential distribution function, the model would be

$$\begin{aligned}P_p &= P(A) + P(B) - P(A)P(B) \\ &= \exp(-t_A/\theta_A) + \exp(-t_B/\theta_B) - \{[\exp(-t_A/\theta_A)][\exp(-t_B/\theta_B)]\}\end{aligned}$$

in failure rate and MTBF terms, the model would be

$$\theta_s = 1/\lambda_A + 1/\lambda_B - [1/(\lambda_A + \lambda_B)] \qquad \text{and} \qquad \lambda_s = 1/\theta_s$$

Quantify the model to predict the reliability of the product. This part may be difficult. Although the other steps are important, they may go for naught if the data used in the prediction is grossly inaccurate (the "garbage-in–garbage-out" syndrome).

Fortunately there are standard handbooks which contain failure rates for most parts and components used in typical products. However, these handbooks suffer one serious limitation—they are almost exclusively oriented toward military products. For example, one of the more popular handbooks, MIL-HDBK-217: "Reliability Stress and Failure Rate Data for Electronic Equipment," contains failure rates for parts (e.g., fixed film resistors) that are qualified to military standards such as

MIL-R-22684. For military standard parts, failure rate determination is pretty straightforward. But we may not be using military parts— they may be commercial parts which haven't been subjected to the qualification steps for the military standards parts.) If this is the case, there are two choices: (1) use failure rates that are directly related to the particular part or (2) modify the military standard part failure rates to account for the fact that they are commercial grade. Of these choices, the former is preferable, since the latter is very judgmental (viz., how much worse is a commercial part than a military standard part?). It should be noted that for some part types in MIL-HDBK-217 (e.g., integrated circuits) an attempt has been made to provide failure rate adjustment factors for commercial equivalents of military standard parts. In any case the caution here is clear: When doing a reliability prediction, failure rates should not be applied blindly.

Since MIL-HDBK-217 is considered the "bible" for many reliability predictions, both for failure rate data and methodology, further insight into this document is provided before going into the process of quantification via example. In using MIL-HDBK-217 for reliability prediction of electronic equipment, there are basically two techniques that can be employed: the parts stress analysis prediction method and the parts count prediction method. The first technique is used when the design has proceeded to a sufficiently detailed stage that there is reasonable knowledge about the parts to be used and the stresses these parts will be subjected to under use conditions. The latter stresses can be assumed, calculated, or measured. The second technique is used during the conceptual or early design stages when detailed design data are not available.

The parts stress analysis prediction method utilizes specific part failure rate models. For the various part types, these models include base failure rates (generally as a function of electrical and thermal stresses) which are adjusted by common factors, as well as by factors which are peculiar to a specific type. The common adjustment factors account for the influence of environmental factors other than temperature (e.g., application in a benign ground environment or hostile aircraft environment) and for the effects of different quality levels (e.g., procured under military specification quality provisions or commercial requirements). The part-type-specific adjustment factors include consideration of important factors influencing failure rate, such as voltage stress and temperature for discrete semiconductor devices; resistance for resistors; capacitance for capacitors; number of brushes for rotating devices; actuation rates and contact loads for relays and switches; and mating and unmating rates for connectors. In the case of microelectronic devices, more expansive failure rate models are used to include consideration of additional factors such as temperature acceleration, voltage derating, package complexity, device learning (e.g., new device in initial production), and programming technique (if applicable). A typical part failure rate model (extracted from MIL-HDBK-217) associated with the parts stress analysis method is given below for read-only memories (ROMs) and programmable read-only memories (PROMS):

$$\lambda_p = \Pi_Q[C_1\Pi_T\Pi_V\Pi_{PT} + (C_2 + C_3)\Pi_E]\Pi_L$$

where $\quad\lambda_p$ – device failure rate in failures per 10^6 h

Π_Q = quality factor

Π_T = temperature acceleration factor, based on technology

Π_V = voltage derating stress factor

Π_{PT} = ROM and PROM programming technique factor

Π_E = application environment factor

C_1 and C_2 = device complexity failure rates based upon bit count

C_3 = package complexity failure rate

Π_L = device learning factor

By contrast, the parts count prediction method is a much simpler-procedure. Generic (or average) failure rates are provided for the various part types. These failure rates are then simply adjusted for factors such as device learning, parts quality, and application environment.

With the data at hand and the prediction technique selected, the process of quantification can begin. Again going back to our simple two-elements-in-series case, assume that these two black boxes, A and B, are in the early design stage and consist of the parts shown in Table 9.1. Because of the level of design detail available at this time, the parts count procedure and failure rate data of MIL-HDBK-217 are used. As shown in this table, for each part type, a quantity and generic failure rate is given; for some parts, additional factors are provided for parts quality and device learning. All generic part failure rates are related to usage in a ground benign environment. A total failure rate is deter mined for each part type as the product of the quantity, generic failure rate, and adjusted by parts quality and device learning factors, as appropriate. The failure rate for each element is obtained simply by summing the total part failure rates. The MTBF is derived by taking the reciprocal of the total element failure rate. Thus, the overall MTBF of the two-series element configuration is 1705 h.

Expressed in probabilistic terms (assuming that each "box" has to operate successfully for a 24-h day), the probability of "no-failure" for this configuration is 0.986. Here is a point that might interest the reader: If we have a black box that has an MTBF (say 24 h) equal to its required "mission" time (again say 24 h) and, assuming an exponential failure distribution, the probability of "no-failure" for this box over the required operating time is only $e^{-t/\theta} = e^{-24/24} = e^{-1.0} = 0.37$!

TABLE 9.1 Sample Reliability Prediction

Part type	Quantity, n	Generic failure rate, λ_G (failures/ 10^6 h)	Quality factor, Π_Q	Learning factor, Π_L	$n\lambda_G\Pi_Q\Pi_L$
		Element A			
Bipolar random logic IC (101–500 gates)	180	0.017	8	10	244.8
Bipolar RAM IC (321–576 bits)	40	0.034	8	10	108.8
MOS RAM IC (577–1120 bits)	32	0.024	8	10	61.44
Silicon NPN transistor	2	0.0025	1	—	0.005
Fixed film resistor, RLR	41	0.0012	1.5	—	0.0738
Fixed film resistor, RNR	117	0.0014	1.5	—	0.2457
Fixed ceramic capacitor, CKR	124	0.0036	1.5	—	0.6696
Fixed aluminum oxide capacitor, CU	12	0.074	1.5	—	1.332
Low-pulse transformer	1	0.0029	1.5	—	0.0029
Printed wiring board connector	4	0.0027	—	—	0.0108
	553				$\lambda_A = 417.4 \times 10^{-6}$ MTBF = 2396 h
		Element B			
MOS random logic IC (101–500 gates)	120	0.012	8	10	115.2
MOS RAM IC (321–576 bits)	60	0.01	8	10	48.0
Bipolar ROM IC (321–576 bits)	10	0.005	8	10	4.0
Silicon NPN transistor	2	0.0025	1	—	0.005
Silicon general purpose diode	2	0.0068	1	—	0.00136
Fixed film resistor, RLR	36	0.0012	1.5	—	0.0648
Fixed film resistor, RNR	120	0.0014	1.5	—	0.252
Fixed ceramic capacitor, CKR	102	0.036	1.5	—	0.5508
Fixed aluminum oxide capacitor, CU	8	0.074	1.5	—	0.888
Low pulse transformer	1	0.0029	—	—	0.0029
Printed wiring board connector	3	0.0027	—	—	0.0081
	464				$\lambda_B = 169.0 \times 10^{-6}$ MTBF = 5917 h

$$\lambda_s = \lambda_A + \lambda_B = (417.4 + 169.0) \times 10^{-6} = 586.4 \times 10^{-6} \text{ failures/h}$$
$$\text{MTBF}_s = 1/\lambda_s = 1/(586.4 \times 10^{-6}) = 1705 \text{ h}$$

If we take these prediction results for elements A and B, and plug them into the math model for these elements arranged in parallel, rather than series, we obtain the following results:

$$\text{MTBF} = 6608 \text{ h and P(24 h)} \cong 0.996$$

As can be seen, both the MTBF and success probability are significantly higher for the parallel arrangement than for the series arrangement. This is not by chance. Redundancy does improve reliability (although not without the penalties of increased cost, weight, volume, power, etc.).

Maintainability. *Define the product in as much detail as possible.* This was also the first step for reliability prediction. Useful information for product definition purposes would include (a) functional block diagrams, (b) mechanical layout drawings, (c) product "packaging" approach (i.e., how are the elements of the product put together?), (d) product maintenance instructions, and (e) fault diagnostics techniques.

Establish the maintenance philosophy to be applied to keep the product working. Some questions that have to be answered at this step include the following:

What level of skills is required of the maintenance people?

What kinds of tools, equipment, and facilities are needed for maintenance?

Which kinds of maintenance actions will be accomplished by the user and which will be done elsewhere, e.g., back at the factory?

How does one determine what needs to be fixed?

How does one remove and replace the faulty product element?

Select an appropriate mathematical model which translates the repair functions into terms that can be quantified. This step is analogous to the one applied for reliability prediction. However, whereas the reliability prediction model can generally turn out to be quite complex, the maintainability prediction model is usually quite simple. The model is related to the way the maintenance action times are expected to be distributed, e.g., as in a log-normal distribution.

Once again, as in the case of the reliability prediction model, the maintainability model is developed to be consistent with the specified requirements. As a rule, maintainability requirements are specified as average or mean times; e.g., mean repair time and mean corrective maintenance time. Some of the more popular maintainability prediction models are shown in Table 9.2. From this table, one thing is obvi-

TABLE 9.2   Typical Maintainability Prediction Models

Equipment Repair Time (ERT):

Repair times follow a normal distribution:

$$ERT = \frac{\Sigma(\lambda\, R_t)}{\Sigma\lambda}$$

Repair times follow an exponential distribution:

$$ERT = \frac{\Sigma(\lambda R_t)}{\Sigma\lambda} \times 0.69$$

Repair times follow a log-normal distribution:

$$ERT = \frac{\Sigma(\lambda R_t)}{\Sigma\lambda} \times 0.45$$

Mean Corrective Maintenance Time:

$$\overline{M}_{ct} = \frac{\Sigma(\lambda M_{ct})}{\Sigma\lambda}$$

Mean Preventive Maintenance Time:

$$\overline{M}_{pt} = \frac{\Sigma(f M_{pt})}{\Sigma f}$$

Notation

R_t = repair time required to perform a corrective maintenance action

λ = element failure rates

M_{ct} = corrective maintenance task time

M_{pt} = preventive maintenance task time

f = frequency of preventive maintenance actions

ous; the models for repair/corrective maintenance consider how often the product fails. Hence, maintainability predictions tie in closely with the product's reliability.

Quantify the model to predict the maintainability of the product. In order to quantify the maintainability prediction model—in the case of repair/corrective maintenance—we need two kinds of inputs: reliability prediction inputs and estimates of maintenance action time. The former types of inputs were described in the previous section and are not repeated here.

The most logical and elementary way to arrive at maintenance action time estimates is to do what is commonly referred to as a time-line analysis. In this type of analysis, the approach is aimed at (1) examining the most likely maintenance actions (e.g., correcting a generator failure in an automobile); (2) breaking this maintenance action into its major task elements (e.g., detecting that there is an electrical failure via an "idiot" light, isolating the failure to the generator via electrical tests, removing the failed generator and replacing it with a new or rebuilt generator, and verifying that this replacement corrected the problem); and (3) making time estimates for these task elements. The latter amounts to a time study and can be done solely in an analytical manner or by taking a mock-up of the product and putting a stopwatch to the possible maintenance actions.

The time-line analysis procedure is described in a military document, MIL-HDBK-472, "Maintainability Prediction." In a simpler procedure, the analyst evaluates such factors as product maintainability design features (e.g., maintenance personnel skill levels required) via a checklist scoring approach and uses the results to arrive at estimated maintainability results.

As an example of a maintainability prediction, assume that we need to estimate the average corrective maintenance time for the same two-element configuration for which we did the reliability prediction in the previous section. Based on an analysis of the design and using mock-ups where available, we can develop corrective maintenance times for the following tasks:

1. *Localization.* The time required to determine the location of a failure to the extent possible without using accessory test equipment (e.g., using diagnostics software).

2. *Isolation.* The time required to determine the location of the failure using accessory test equipment.

3. *Disassembly.* The time required to disassemble the product to the extent necessary to gain access to the item to be replaced.

4. *Interchange.* The time required to remove the defective item and install the replacement.

5. *Reassembly.* The time required to close and reassemble the product after the replacement has been made.

6. *Alignment.* The time required to perform any alignment, minimum tests, and/or adjustment made necessary by the repair action.

7. *Checkout.* The time required to perform the minimum checks or tests required to verify that the product has been restored to a satisfactory operating condition.

Using the failure rate estimates derived earlier (see Table 9.1) and corrective maintenance times developed hypothetically for the above tasks, the mean corrective maintenance time can now be determined for the two-series element configuration. As shown in Table 9.3, the total corrective maintenance task time for each element is weighted by the expected failure frequency (failure rate); the maintainability prediction model given in Table 9.2 is used to arrive at a mean corrective maintenance time M_{ct} of 34 min for the configuration.

The results of reliability and maintainability predictions can be used in a variety of ways: (1) to estimate the capability of a product, (2) to help establish requirements through an allocation process, (3) to help identify potential weak links and problem areas requiring improvement, and (4) to aid in assessing the existence of a risk condition in the product.

Table 9.3   Sample Maintainability Prediction[*]

Element	Failure rate, λ	Localization time	Isolation time	Disassembly time	Interchange time	Reassembly time	Alignment time	Checkout time	M_{ct}	λM_{ct}
A	417.4	0.069	0.071	0.094	0.051	0.134	0.21	0.118	0.558	232.91
B	169.0	0.058	0.043	0.094	0.051	0.134	0.042	0.171	0.593	100.22
	586.4									333.13

$$\bar{M}_{ct} = \frac{\Sigma(\lambda M_{ct})}{\Sigma\lambda} = \frac{333.1}{586.4} = 0.57 \text{ h} \cong 34 \text{ min}$$

*NOTE: 1. All failure rates are in failures per million hours. 2. All corrective maintenance times are in hours.

9.3.2   Allocation

One of the responsibilities of a reliability/maintainability engineer is to provide guidance to the product designer. This guidance comes in various forms, not the least of which are numerical requirements or goals. Let's say we have a black box that has an MTBF requirement of 1000 h and a MTTR requirement of 15 min. Further assume that this box is made up of four major assemblies, each of which is assigned to a separate electrical and mechanical designer. How do each of these designers know they are going in the right direction toward meeting the overall reliability/maintainability requirements?

First, if the reliability/maintainability engineers are doing their job properly, they will provide the designers with a set of qualitative design guidelines—the kind of guidelines that everyone takes for granted (e.g., derate parts to x percent of their rated values, use modular packaging). Second, the reliability/maintainability engineers will furnish the designers with numerical requirements/goals which are derived from the overall requirements. These design targets will serve as guideposts against which measurements can be made during the design phase. These measurements are made by the reliability/maintainability engineers in the form of predictions of reliability/maintainability capability. When compared with the numerical requirements/goals, the prediction results can be used to identify potential reliability/maintainability problem areas. Thus, not only does the allocation tell the designer how difficult the design task is in numerical terms, but it also provides a basis for identifying product areas which require improvement.

The techniques for performing reliability/maintainability allocations can be simple or sophisticated. The simpler techniques presuppose a minimum of product design information, while the more sophisticated methods usually require more detail so that the allocation can be optimized with due consideration to other product parameters (e.g., weight, cost, volume). In a sense, the more sophisticated techniques usually involve optimization approaches in which trade-offs are made to maximize the system parameters. The problem here is that while the sophisticated allocation process goes on, the designer must wait. Thus, one should get the design targets to the designer early, using simple-minded techniques if necessary. Next, update and revise these allocations as the design effort progresses with an aim toward optimizing all system parameters.

In order to narrow the field of allocation approaches, only two simple techniques will be presented—one each for reliability and maintainability.

Reliability.   Perhaps the simplest technique for reliability allocation is the "equal-apportionment technique." The following steps show how simple this can be.

1. Perform a reliability prediction of the product.
2. Tabulate the prediction results by major product elements consistent with designer responsibility.
3. Establish a scaling factor (the ratio of required to predicted capability).
4. Adjust the predicted element reliability levels using the scaling factor to derive the allocation.

Mathematically, the above process reduces to the following when failure rates are to be allocated:

$$\lambda_i = K\lambda_i$$

where λ_i = allocated element failure rates
   λ_i = predicted element failure rate
   $K = \lambda_s'/\lambda_s$ = scaling factor (note: λ_s' = overall required product failure rate and λ_s = overall predicted product failure rate)

To illustrate, go back to the reliability prediction for the two elements (see Table 9.1). For the two-series element configuration the predicted MTBF was 1705 h. If the MTBF requirement were 2000 h, how would we go about allocating the MTBF goals for elements A and B? Using the simple relationship given above,

$$\lambda_s' = \frac{1}{2000} = 500 \times 10^{-6} \text{ failures per hour}$$
$$\lambda_s = \frac{1}{1705} = 586 \times 10^{-6} \text{ failures per hour}$$
$$K = \frac{500}{586} = 0.853$$
$$\therefore \lambda_A' = (417.4 \times 10^{-6})(0.853)$$
$$\cong 356 \times 10^{-6} \text{ failures per hour or MTBF}_A' = 2809 \text{ h}$$

$$\lambda'_B = (169.0 \times 10^{-6})(0.853)$$
$$\cong 144 \times 10^{-6} \text{ failures per hour or MTBF}'_B = 6944 \text{ h}$$

Thus, the failure rates for each element have to be reduced down to the levels indicated to achieve the overall MTBF requirement of 2000 h. As pointed out earlier, one way to satisfy this requirement is through the use of redundancy. There are other ways that could be used to enhance reliability, e.g., selection of more reliable parts, reduction of environmental stresses, or reduction in complexity.

Obviously, the above technique is not the best one for allocation of reliability requirements or goals. It has at least one serious limitation: It assumes that reliability improvements are equally proportional and feasible in all elements. Furthermore, it doesn't fully or adequately consider a product that has redundancy; it is more appropriate for a product whose elements are in series from a reliability viewpoint. These shortcomings notwithstanding, even this simple method provides useful design targets which can later be updated, revised, and optimized using more sophisticated techniques.

Maintainability.   An equally simplistic technique for maintainability allocation is as follows:

1. Perform a reliability prediction and estimate the failure rate (λ_i) of the major product elements using the techniques described earlier.

2. Make first-cut corrective maintenance time estimates (M_{ct_i}) for each major product element.

3. Determine the failure contribution of each major element (C_i) to the expected total failures as:

$$C_i = n_i \lambda_i$$

4. Determine the predicted contribution of each element to corrective maintenance time as the product of C_i times M_{ct} the predicted corrective maintenance time.

5. Determine the scaling factor to be used to adjust the predicted element corrective maintenance times as:

$$K = \frac{M'_{ct_i} \Sigma C_i}{\Sigma C_i M_{ct_i}}$$

6. Adjust the predicted element corrective maintenance times to arrive at the element allocations:

$$M'_{ct_i} = K_i M_{ct_i}$$

As an example, go back once again to our simple two-series element configuration. We can make use of the reliability and maintainability prediction results given in Tables 9.1 and 9.3, respectively. Assume that the overall requirement for this configuration is 0.5 h. The overall-predicted corrected maintenance time $(\bar{M}_{ct})$ for this configuration is 0.57 h. Using the allocation procedure described above, we would end up with a table that looks like Table 9.4. The calculated K factor is used to scale the M_{ct} values down to a level M'_{ct} such that the overall $\bar{M}_{ct}$ requirement is 0.5 h. The next step would be to establish the means for achieving these new M'_{ct} goals (e.g., by incorporating built-in test equipment to reduce the time to localize failures).

The above technique suffers the same kinds of shortcomings as the method for reliability allocation described earlier. In some cases, improvements are not feasible. Therefore, engineering judgment may be necessary to make further adjustments to these allocations. Again though, as preliminary design goals they are valuable and can later be optimized.

At this point, it is important to note that allocations, such as those made for reliability and maintainability, provide a good basis for initiating product trade-offs. A review of these allocations and/or a comparison with predictions of product capability could easily result in the conclusion that you "can't get there from here." For example, if the prediction shows that the allocated maintainability goal for an element is impractical to achieve, it may be necessary or easier to enhance reliability and reduce the probability of failure—thus requiring less maintenance of that element.

TABLE 9.4 Sample Maintainability Allocation

Element	Quantity, n_i	Predicted failure rate $\lambda_i \times 10^{-6}$, failures/h	Failure contribution $C_i \times 10^{-6}$, failures/h	Predicted corrective maintenance time M_{ct}, h	Predicted contribution to corrective maintenance time $C_i M_{ct}$ $(\times 10^{-6})$	Allocated corrective maintenance time $K_i M_{ct} = M'_{ct}$, h
A	1	417.4	417.4	0.558	232.91	0.49
B	1	169.0	169.0	0.593	100.22	0.52
Totals		586.4			333.13	

$$\bar{M}_{ct} = \frac{\Sigma C_i M_{ct}}{\Sigma \lambda_i} = \frac{333.1 \times 10^{-6}}{586.4 \times 10^{-4}} = 0.57$$

$$M'_{ct} \Sigma C_i = (\Sigma n_i \lambda_i)(\bar{M}'_{ct}) = (586.4 \times 10^{-6})(0.5) = 293.2 \times 10^{-6} \text{ failures per hour}$$

$$K = \frac{M'_{ct} \Sigma C_i}{\Sigma C_i M_{ct_i}} = \frac{293.2 \times 10^{-6}}{333.1 \times 10^{-6}} = 0.88$$

The principal value of the allocation process is for establishing design goals or requirements. However, when compared with the prediction results, these allocations can be used to detect potential design weaknesses which can be earmarked for improvement.

9.3.3 Assessment

The analytical tools described so far are related to design support activities. Presumably they have steered you in the right direction toward meeting your requirements. You have now transferred your product design from paper (i.e., schematics, drawings, etc.) to hardware. In-plant tests are run on the product, and then it is delivered to the customer. One way to measure the capability of the product is to assess how it actually performs.

In such an assessment, we must realize one thing—we are only looking at a sample of the product's capabilities. For example, we might have only one unit of a product and assess its capabilities over only a portion of its lifetime; or we might take some units of a product and assess their capability over their whole lifetime. In either case, we are only looking at a sample—in the first case, a sample of the product's operational lifetime, and in the second, a sample of the product population. When sampling we have uncertainty. Since we have not been able to look at all of the product, over all of its lifetime, we can only analyze what we know about the product, in terms of data, and extrapolate from these data. Fortunately, we can apply some estimation techniques and utilize statistical inference theory to help.

Consider the firecracker example already discussed. As you recall, we wanted to know how good the lot of firecrackers were. But if we tested them all, we wouldn't have any to sell, so we drew a sample of 100 and fired them off. We found that 95 out of 100 fired off. We asked ourselves the question, "How reliable are the firecrackers?" Well, the so-called point estimate was determined simply as

$$\text{Point estimate} = \frac{\text{no. successes}}{\text{total trials}} = \frac{95}{100} = 0.9$$

Thus, the point estimate of the reliability of these one-shot devices was 0.95 or 95 percent. Similarly, if we wanted to measure the capability of a product in the time-domain, we could calculate a point estimate. Assume that we wanted to assess the reliability capability of such a product, putting a sample of 25 repairable units on test for 100 h. Five units experienced failures during the test. For this set of data, the point estimate or MTBF ($\hat{\theta}$) would be determined as

$$\text{Point estimate} = \hat{\theta} = \frac{\text{total time}}{\text{total failures}} = \frac{25{,}000}{5} = 5000\,\text{h}$$

Although these point estimates give us an idea of the product's capability, we need to look a little harder at the data to get a better indication. Hence, we resort to the use of confidence limits and statements. Confidence limits come in two varieties—one-sided and two-sided. In the former case, we say that the "true" capability of the product (had we been able to measure 100 percent of the product over all of its lifetime) is no worse (or better) than a single limit; in the latter case, we say that the true capability of the product lies within an interval which is bounded by a lower and an upper limit. Since the determination of these limits is related to some statistical confidence level, we can make a confidence statement about the results. Thus, we could say that we have 95 percent probability that the true MTBF is equal to or greater than χ or lies between χ_L and χ_U. Generally speaking, the greater the sample size, the greater the degree of confidence that one can place in the results; i.e., the limits are tighter. Conversely, the greater the degree of confidence desired, the wider the limits are.

Since the mathematical relationship used for the determination of these confidence limits is dependent on the use of an appropriate distribution function, we again resort to some of the fundamentals brought out earlier in this chapter. Upon selection of an appropriate function (either through experience or through use of statistical tests beyond the scope of this book), mathematical formulas of the type given in Table 9.5 can be used to calculate the confidence limits at a given confidence level.

For our time-oriented example then, we could generate a set of one- and two-sided confidence limits, at the 90 percent confidence level. Obviously, similar limits could be determined for other confidence levels and other device types. It should be noted that the lower 90 percent one-sided limit is *not* equal to the lower 90 percent confidence limit, since the former is equivalent to the 10 percent point on the distribution curve and the latter is equivalent to the 5 percent point.

Point estimate	Lower 90% one-sided limit	90% two-sided limits	
		Lower	Upper
5000 h	2696 h	2378 h	9568 h

Assessment results of the type tabulated here can be used to measure conformance with the requirements. However, by nature of the process

TABLE 9.5 Typical Confidence Limit Formulas

Exponential Distribution

Point Estimate: $\hat{\theta} = \dfrac{T}{r}$

Lower Confidence Limit: $\theta_L = \dfrac{2T}{\chi_a^2}$

Upper Confidence Limit: $\theta_U = \dfrac{2T}{\chi_b^2}$

One-Sided Lower Confidence Limit: $\theta_{LS} = \dfrac{1}{\chi_c^2}$

Binomial Distribution

Point Estimate: $\hat{R} = \dfrac{S}{N}$

Lower Confidence Limit: $R_L = \dfrac{1}{1 + \left(\dfrac{r+1}{N-r}\right)F_a}$

Upper Confidence Limit: $R_U = \dfrac{1}{1 + \left(\dfrac{r+1}{N-r}\right)F_b}$

One-Sided Lower Confidence Limit: $R_{LS} = \dfrac{1}{1 + \left(\dfrac{r+1}{N-r}\right)F_c}$

Notation

T = total operating time
N = total number of trials
r = total number of failures
χ^2 = chi-square distribution value associated with $(2r + 2)$ degrees of freedom at the given probability level
F = F distribution value using r and N as entering arguments at the given probability level
$1 - \alpha$ = desired confidence level
a = probability level $\alpha/2$
b = probability level $(1 - \alpha/2)$
c = probability level (α)
S = number of successful trials
R = probability of success
θ = mean-time-between-failures

involved—evaluation of sample results—we are also predicting the capability of the product.

9.3.4 Failure modes, effects, and criticality analysis

Up until now, quantitative analytical tools have been described. Yet, one of the most powerful analytical design tools is a qualitative one—the failure modes, effects, and criticality analysis (FMECA). As noted in the chapter on design, the FMECA provides an orderly approach toward evaluating a product design to determine (a) the ways in which the product can fail, (b) the consequences of these failure modes, and (c) the seriousness of such failures on the reliability of the product. Although the FMECA is primarily a qualitative analytical tool, a quantitative flavor is usually imparted because the failure frequency of each mode is also considered in the analysis. The aim of the FMECA is to identify those areas of the product design that need to be improved—generally speaking, any areas that contribute failure modes that (a) have a relatively high frequency of occurrence, (b) have a critical impact on the product's ability to do what it's supposed to do, and (c) have no existing or compensatory design features to circumvent the failure consequences.

As an example of an FMECA, take a look at an automobile starter in terms of several gross failure modes. Table 9.6 shows a typical tabular format that could be employed in such an analysis. For each gross failure mode, an estimate is made of its relative failure frequency. In our example, this estimate is provided in qualitative terms; however, a quantitative estimate (e.g., in terms of probability of occurrence or failure rate) could easily be substituted. The effect and criticality of each failure mode is also examined. The criticality classification is subjective and dependent on the "mission" of the product. In our case, the product under analysis is an automobile and its mission is to get us where we want to go. If the automobile won't start, it could be critical. In actual practice, such a FMECA would be further expanded to determine the cause(s) of each failure mode and to identify existing or potential design compensatory features (e.g., on some smaller older foreign automobiles, a hand crank can be used).

The FMECA can be performed at various levels of the product, including the major assembly or the piece-part level. Selection of an analysis level is usually dependent on cost—the lower the level at which the FMECA is done, the more costly. However, the lower the analysis level, the greater the chances of finding seemingly innocent failure modes. Thus, trade-offs have to be made by both the producer and the consumer to establish a cost-effective analysis level.

TABLE 9.6 Sample Failure Modes, Effects, and Criticality Analysis (FMECA)

Item	Gross Failure Modes	Relative Failure Frequency	Failure Effects	Failure Criticality
Automobile starter	(a) Burned starter solenoid switch contacts	High	Slow engine cranking speed	Major-automobile may not start depending on degree of contact deterioration
	(b) Defective starting motor	Medium	Slow engine cranking speed; starter will not run	Major/critical-automobile may not start depending on type of motor defect
	(c) Bent armature shaft or damaged drive mechanism	Low	Starter engages, but will not crank	Critical-automobile will not start
	(d) Faulty armature or field	Low	Starter engages, but will not crank	Critical-automobile will not start
	(e) Shorted or open starter circuit	Medium	Starter will not run	Critical-automobile will not start
	(f) Defective solenoid switch	High	Starter will not run	Critical-automobile will not start

In terms of actually doing the analysis, one needs a product design and an understanding of the way in which it functions. Thus, the FMECA can be initiated in the early design phase and conducted at a higher assembly level (e.g., subsystem), and then later as the design becomes more definitive, perhaps carried down to lower levels (e.g., piece part). The general approach to the FMECA is in "bottoms-up" fashion—starting at the lowest level of the product at which the analysis is to be done and proceeding upward to the end item, i.e., part, component, assembly, etc. The results are invaluable in identifying reliability weak links so that improvements can be initiated.

Hazard analysis

A qualitative analytical tool available to the systems safety engineer is the hazard analysis. The aim of a hazard analysis is to provide a technical assessment of the relative safety of a product design. In this day of product liability suits, the implementation of such analyses is becoming increasingly important to the producer.

The hazard analysis effort usually begins with a preliminary evaluation of the product, the output of which is used to develop a set of safety criteria and to identify any gross safety problem areas. For example, the preliminary hazard analysis would be directed toward ensuring the compatibility, from a safety viewpoint, of materials used in the product. This stage of the hazard analysis is analogous to the FMECA which is performed as part of the reliability activity. However, whereas the FMECA looks at the modes and effects of failures on equipment reliability, the hazard analysis examines failure modes and their effects on personnel *and* product safety during operation and maintenance. For large equipment level evaluation, the hazard analysis would be further extended to include integration and interfaces within the context of an overall system.

A later stage of the hazard analysis is the so-called operating hazard analysis. It is intended to be used to determine safety requirements for personnel, operation, maintenance, procedures, testing, transportation, storage, and training. The outputs of the engineering design and test programs are used to develop safety considerations such as warning and caution statements in product operating/maintenance procedures. Obviously the scope and depth of such analyses depends greatly on the type of product, but the purpose is the same—eliminate, minimize, or control safety hazards.

Tables 9.7 and 9.8 show examples of typical hazard analysis worksheets. The first table presents an abbreviated fault hazard analysis (FHA), one of the design-oriented types of hazard analyses. In this first example, we are looking at a missile component, a booster igniter, for

TABLE 9.7 Sample Fault Hazard Analysis (FHA)

Component	Failure Mode	Probability of Occurrence	System Operational Mode	Effect of Primary Component Failure on End item
Rocket igniter	Igniter fires without required time delay	1×10^{-6}	Rocket firing	Premature rocket launching severe equipment damage will be sustained and possible injury or death to personnel.

safety hazards. The FMECA and the FHA are similar in that failure modes are evaluated for frequency of occurrence, effect, criticality, etc. However, we are interested in the safety aspects rather than reliability. The second table provides a brief version of an operating hazard analysis (OHA) that is employed during the validation phase to prevent the need for costly product modification or redesign after hardwaredesign has been completed. Both types of analyses serve to provide a basis for establishing safety controls and improvements.

Thus, the overall hazard analysis effort can be used to (a) establish product safety requirements, (b) assess product safety capability in qualitative terms, (c) detect potential safety problem areas which require improvement, and (d) evaluate product safety risk.

9.3.5 Fault tree analysis

The fault tree analysis (FTA) is a powerful tool for use in evaluating the safety consequences of product anomalies (e.g., in the nuclear power industry) in quantitative terms. (It should be noted that an FTA can also be accomplished in qualitative terms, with the output limited to the identification of hazardous events.) A fault tree is a graphical representation related to a particular product anomaly; e.g., failure of an overload protection device. The FTA is a top-down type of analysis, in which each of the events which contribute to a particular anomaly are

Factors That May Cause Secondary Component Failure	Upstream Components that May Command Undesired Event	Hazard Duration Time	Hazard Classification	Recommended Hazard Control
(1) Personnel error—improper igniter installed (2) Thermal energy causes incorrect igniter firing	Delay gate contact failed in closed position	9 seconds	Critical	(1) All igniters must be properly inspected to ensure correct igniter has been installed by personnel. (2) Redesign firing circuit to incorporate two delay gate contacts in series, rather than a simple gate contact.

evaluated in both quantitative and qualitative terms. Stated another way, the fault tree shows the cause-effect relationship between the top undesired event and various contributing events, by providing a logical statement of the cumulative effect of faults within a product.

In order to perform an FTA, the systems safety analyst needs to go through the following steps:

1. Define the product in terms of the product faults to be evaluated.

2. Construct the fault tree in terms of the subevents that contribute to the occurrence of the faults (top events) to be analyzed.

3. Evaluate the various fault tree subevents in qualitative terms to gain insight into the unique modes of product faults.

4. Quantify the fault tree events, considering the cumulative effects of anomalies within the product to enable complete analysis of the probability of occurrence of the top event—product failure.

Step 4 requires the application of combinatorial logic network analysis. In this type of analysis, the failure effects of the various subevents are tracked through logical AND or OR gates until they result in a particular product fault. A typical fault tree, expressed in probabilistic terms, is shown in Figure 9.4. It is in such analyses that a knowledge of probability theory—which was briefly touched upon earlier in

TABLE 9.8   Sample Operating Hazard Analysis (OHA)

Function	Operation	Effect (Normal)	Effect (Hazard)	Hazard Category	Safety Features Recommended Controls
Fire switch (located on Rocket Control Panel)	Activate rocket firing circuit under emergency conditions (e.g., rocket destruct)	Normal rocket firing	Fire switch is physically located too close to "Arming Switch" on panel; due to its close proximity, the switch can be inadvertently activated when arming	Critical	Relocate fire switch a sufficient distance from arming switch; also provide a protective cover over fire switch

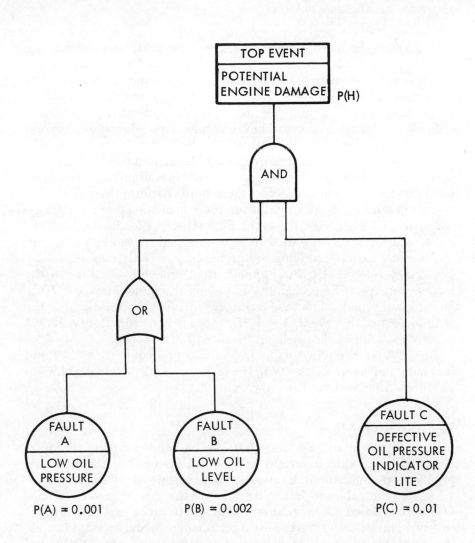

$$P(H) = [P(A) + P(B) - P(A)\ P(B)\ \ P(C)]$$

$$= [0.001 + 0.002 - (0.001)\ (0.002)]\ 0.01$$

$$= (0.002998)\ (0.01) = 0.00002998$$

Figure 9.4 Sample fault tree analysis (FTA).

this chapter—becomes important. After the fault tree events are expressed in probabilistic terms, they are quantified using actual experience data or estimates, in much the same manner as a reliability prediction.

As an example of a fault tree analysis (FTA), consider an undesirable top event such as potential automobile engine damage. The ways in which such damage can occur will be limited to a combination of three gross faults; low oil pressure, low oil level, and defective oil pressure indicator light. That is, if there is *either* low oil pressure *or* low oil level *and* the driver fails to detect these faults because there is a defective oil pressure indicator light, then there could be potential damage to the engine. Figure 9.4 shows the FTA structure for this evaluation. Appropriate symbols are used to represent logical AND and OR gates, and a math model describing the probability of occurrence of the top event is developed. Utilizing hypothetical probability values in the model, we expect the probability of potential engine damage $P(H)$ to be extremely small—less than 3 chances in 100,000. Or looking at the complement of this event, $1 - P(H)$, there is approximately a 99.997 percent probability that such an event will not happen.

The FTA offers obvious value in evaluating product capability, but it is more important in assessing the product for identification of high-risk areas that will require improvement.

9.3.6 Control charts

Up until now, analytical tools that can be applied during the build phase from a product assurance viewpoint have been ignored. When a product is manufactured, there are many reasons why its quality may vary. However, the causes can be put into the two general categories of *chance* and *real*. *Chance* causes of variation can be expected to occur in any manufacturing process—to a certain degree and hopefully exhibiting a stable pattern. The problem lies in being able to identify and correct the *real* causes of process variation.

Before examining the subject of control charts, we must first review the way that the integrity of a product is evaluated from a quality viewpoint. Depending on the quality characteristic or level that we are interested in controlling, we can apply control charts to the control of *attributes* or *variables*. Attributes relate to how many good versus how many bad. For example, a quality specification for resistors might state that only two rejects are permitted for each 100 resistors tested. On the other hand, variables are concerned with measurements that reflect product quality. For example, the quality specification for resistors could be stated as a particular ohmic value (e.g., 100 Ω), with a given tolerance range (e.g., 100 Ω plus or minus 5 percent). Obviously, these two quality criteria are often combined for control of product quality. The quality requirement could be specified as "100 Ω ± 5 percent, with two rejects allowed for each 100 resistors tested."

Control charts translate attributes and variables data into a form which can be used to help control product quality. For attributes data, there are two types of control charts which can be employed—one for percent defective ($\bar{p}$) and the other for the quantity of defects ($\bar{c}$ or $\bar{\mu}$) per unit. For variable data, there are three control charts which are typically used—one for averages ($\bar{X}$), another for ranges ($\bar{R}$), and the last for standard deviation (σ). The first gives a view of the product's central tendency for a given measurable characteristic, while the last two provide insight into the variability of the measurements for that characteristic.

Table 9.9 provides a summary of the typical types of control charts which can be applied, the factors they are aimed at controlling, and the parameters of interest for each type. Before delving into some of the unique features of each, let us review some of the common aspects. Each has an average value, a standard value, and control limits. How do these parameters fit together for process control purposes? Quality control practitioners speak in terms of a constant-cause system, i.e., a system in which inspection results vary over some range but are stable. If a process is a constant-cause system, it is in fact stable and in statistical control. And what better way to make this determination than through the use of the control chart?

The averages values shown in Table 9.9 are based on past experience. Having determined this average value of quality performance, we can establish a target or standard value against which future performance can be tracked. This standard value can be set equal to or different than the average experienced value, depending on whether the process is in control or on whether the standard value is a reasonable objective. The control limits provide a way of determining whether the process is in or out of control. A quick look at the factors in each control limit equation reveals that they have commonality in that the control limits are about the average (or standard value); they are either directly or indirectly a function of the sample size n; and the attributes control limits contain a "3," which relates to the three-sigma points in a normal distribution curve (also called the gaussian curve). Since we are interested in comparing the quality of the product relative to past experience or a desired standard, we set the control limits about the average or standard value. The sample size comes into play because the degree of certainty in setting the limits is a function of the sample size—the greater the sample size n, the tighter the limits and vice versa for a fixed average or standard value. For the attributes control charts, three-sigma limits are commonly used because they offer a reasonably high

TABLE 9.9 Typical Control Chart Parameters

Control Chart Type	Control Factor	Average Experienced Value	Standard Value	Control Limits
Attributes	Fraction Defective[a]	$\bar{\rho} = \dfrac{\text{No. of Defectives}}{\text{No. Inspected}}$	ρ'	$\rho \pm \dfrac{3\sqrt{\bar{\rho}(1-\bar{\rho})}}{\sqrt{n}}$
	Defects (per Single Unit)	$\bar{c} = \dfrac{\text{No. of Defects}}{\text{No. Inspected}}$	c'	$\bar{c} \pm 3\sqrt{\bar{c}}$
	Defects (per Multiple Units)	$\bar{\mu} = \dfrac{\text{No. of Defects}}{\text{No. Inspected}}$	μ'	$\bar{\mu} \pm \dfrac{3\sqrt{\bar{\mu}}}{\sqrt{n}}$
Variables[e]	Averages	$\bar{\bar{X}} = \dfrac{\text{Sum of Subgroup Averages}}{\text{No. of Subgroups}}$[b]	$\bar{X}'$	$\bar{\bar{X}} \pm A_1 \bar{\sigma}$, or $\pm A_2 \bar{R}$
	Standard Deviations	$\bar{\sigma} = \dfrac{\text{Sum of Subgroup Standard Deviations}}{\text{No. of Subgroups}}$	$\bar{\sigma}'$	$\bar{\sigma} + B_3\,\bar{\sigma}$ and $\bar{\sigma} - B_4\,\bar{\sigma}$
	Ranges	$\bar{R} = \dfrac{\text{Sum of Subgroup Ranges}}{\text{No. of Subgroups}}$	$\bar{R}'$	$\bar{R} + D_3\bar{R}$, and $\bar{R} - D_4\bar{R}$

[a]For percent defective, take 100 times the fraction defective

[b]A subgroup is a division of sample observations

[e]Values of the constants A_1, A_2, B_3, B_4, D_3, and D_4 may be found in any book on statistical quality control. Calculation of the numerators of the variables equations related to "Average Experienced Value" may be obtained as follows:

$$\text{Subgroup average} = \frac{\text{sum of observations in a given subgroup}}{\text{no. of sample observations in subgroup}}$$

$$\text{Subgroup standard deviation} = \sqrt{\frac{\text{sum (deviations of observations from average for a given subgroup)}^2}{\text{no. of sample observations in subgroup}}}$$

Subgroup range = difference between highest and lowest values of the observations in a given subgroup

Note: A more detailed treatment of the above relationships can be found in any book on statistical quality control.

probability (.9973) that any observation falling outside is due to some assignable cause—not just chance (see Table 9.10).

Once the process is found to be in or out of control, based on the established limits, a course of action must be decided upon. If the process is in statistical control, it is generally left alone. On the other hand, if it is out of control, we have to hunt for an assignable cause(s) and determine why we do not have a constant-cause system.

Examples of the two types of control charts are shown in Figures 9.5 and 9.6 for attributes and variables, respectively. In each case, a set of hypothetical data (as shown in Tables 9.11 and 9.12) is used to calculate the mean of the process being evaluated, establish a standard value for the process, and determine the upper and lower control limits (UCL and LCL, respectively). In the first example, the number of mechanical defects found on each radio inspected is tabulated for the first 20 radios coming off the assembly line. The calculated mean, $\bar{c} = 7.4$ defects per radio, becomes the standard value, c'; it is also used in the determination of the UCL and LCL. The inspection results of the next 10 radios are also tabulated. The c, c', UCL, and LCL are used to develop the control chart for radio defects. From this chart, it can be seen that the process remains stable until radio unit number 26 comes through the assembly line. The number of defects experienced by this unit is significantly greater than has been observed for the process in the past. Corrective action is necessary to bring the process back into control. In the second example, holes are being punched in a small casting, with samples of four castings being drawn from every 100 castings and measured for hole dimension. The measurements for the first 20 subgroups (containing four sample castings each) is tabulated. A mean, $\bar{X} = 7.4$ thousandths of an inch, is calculated and used as the standard value $\bar{X}'$. The UCL and LCL are established using the calculated range average $\bar{R}$ and a constant A_2 which is a function of the subgroup size, 4. As in the previous example, the inspection results for the next 10 subgroups are tabulated. The control chart for casting measurements is developed using the $\bar{\bar{X}}$, $\bar{X}'$, UCL and LCL. Examination of this chart reveals that the process is quite

TABLE 9.10   Limits and Related Probability Levels in a Normal Distribution

Limits	Probability of Falling Within/Outside These Limits	
	Within	Outside
Mean Value ± 1σ	0.6826	0.3174
Mean Value ± 2σ	0.9546	0.0454
Mean Value ± 3σ	0.9973	0.0027

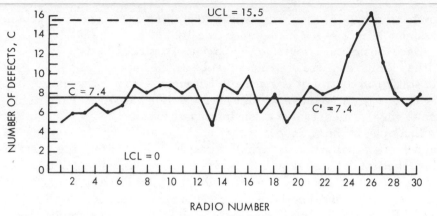

Figure 9.5 Sample attributes control chart: defects per unit.

stable but develops a trend beginning with subgroup number 20 which eventually results in an out-of-control situation with subgroup number 26. Corrective action (e.g., machine recalibration) brings the operation back into control.

Thus, control charts can be used (a) to track the product to ensure conformance with quality requirements and (b) to identify points in the process which need to be corrected and improved from a quality viewpoint.

9.3.7 Design of experiments

Once an abnormal number of manufacturing defects is identified, the next job is to find the cause. One approach toward finding the cause is "brute force"—trying to overwhelm the problem. This approach usually

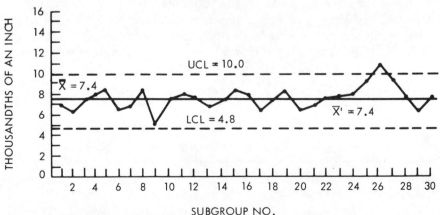

Figure 9.6 Sample variables control chart.

TABLE 9.11   Data and Calculations for Sample Attributes Control Chart (Figure 9.5)

Data	
Radio No.	*No. of Mechanical Defects*
1	5
2	6
3	6
4	7
5	6
6	7
7	9
8	8
9	9
10	9
11	8
12	9
13	5
14	9
15	8
16	10
17	6
18	8
19	5
20	7
	Total 147
21	9
22	8
23	9
24	9
25	12
26	14
27	16
28	8
29	7
30	8

Calculations

$$\bar{c} = \frac{\sum\limits^{n}_{c} }{n} = \frac{147}{20} = 7.4$$

$$UCL = \bar{c} + 3\sqrt{\bar{c}} = 7.4 + 3\sqrt{7.4} = 7.4 + 8.1 = 15.5$$

$$LCL = \bar{c} - 3\sqrt{\bar{c}} = 7.4 - 3\sqrt{7.4} = 7.4 - 8.1 = 0$$

results in the "brute" being overwhelmed without finding the cause. Another approach is the trial-and-error approach. Although somewhat better than the former, this approach is more costly and sometimes also leads to less than satisfactory results. The best approach is to employ a statistically designed experiment—one that is established so that the effects of the variables can be separately analyzed on a statistical basis.

TABLE 9.12 Data and Calculations for Sample Variables Control Chart (Figure 9.6)

			Data			
			Unit Measurement (Thousandths of Inch)			
Subgroup No.	A	B	C	D	$\overline{X}$	R
1	5	6	7	9	6.8	4
2	4	6	8	7	6.3	4
3	11	7	4	8	7.5	7
4	8	7	10	7	8.0	3
5	7	10	8	9	8.5	3
6	6	9	5	6	6.5	3
7	6	7	5	10	7.0	4
8	8	11	9	6	8.5	5
9	6	5	4	5	5.0	2
10	7	7	8	9	7.8	2
11	9	8	7	8	8.0	2
12	11	4	7	9	7.8	7
13	5	6	8	9	7.0	4
14	9	8	7	6	7.5	3
15	8	9	9	8	8.5	1
16	7	7	11	7	8.0	4
17	6	8	6	5	6.3	3
18	8	9	7	6	7.5	2
19	9	8	10	7	8.5	3
20	5	4	8	9	6.5	5
Totals					147.5	71
21	10	5	7	6	7.0	5
22	6	7	8	9	7.5	3
23	9	7	4	11	7.8	7
24	7	11	7	7	8.0	4
25	9	10	10	9	9.5	1
26	12	11	13	11	11.0	2
27	10	10	10	8	9.5	2
28	7	8	6	9	7.5	3
29	6	5	8	7	6.5	3
30	7	7	8	9	7.8	2

Calculations

$$\overline{\overline{X}} = \frac{\sum^n \overline{X}}{n} = \frac{147.5}{20} = 7.4$$

$$\overline{R} = \frac{\sum^n R}{n} = \frac{71}{20} = 3.6$$

$$UCL = \overline{\overline{X}} + A_2\,\overline{R} = 7.4 + (0.73)(3.6) = 10.0$$

$$LCL = \overline{\overline{X}} - A_2\overline{R} = 7.4 - (0.73)(3.6) = 4.8$$

There are a number of experiments that are statistically based. For example, there are randomized block experiments which evaluate the effects of one factor, randomly applied on the specimen under study; there are Latin-squares experiments which evaluate the effects of two factors, randomly applied, on the specimen under study; and there are factorial design experiments which evaluate the effects of a number of factors on the specimen under study. Whatever the approach employed, the aim is essentially the same: (1) to find out which variables have the most significant effects on the cause of defects and (2) to determine what the interactions of these variables are so that they can be best combined to yield a quality product.

Using a properly designed experiment, one can efficiently determine which variables need to be controlled and what improvements need to be initiated.

9.4 Automated Tools

The types of analyses described here lend themselves very nicely to automated approaches. The use of computerized methods offer the obvious advantages of speed and accuracy, and relieve the tedium related to much of the "number-crunching" involved in the quantitative-based techniques. In addition, because of the overwhelming number of up-down and success-fail states involved in some of today's highly complex systems (e.g., telecommunications switching equipment), they can only be realistically analyzed for some product integrity figures of merit (e.g., system availability) by employing computerized approaches.

Computer programs have been developed for just about all of the analytical methods described in this chapter. There are computer programs for probability analysis; reliability and maintainability predictions/allocations; failure modes, effects, and criticality analysis (FMECA); fault tree analysis (FTA); and control charting, to mention just a few. Some of these programs, such as the RADC ORACLE (Optimized Reliability and Component Life Estimator) program, were developed by Department of Defense (DOD) organizations for use on DOD contracts. There are also companies, such as Syscon Corporation of Middletown, Rhode Island, and Sunnyvale, California, which provide computer-aided products and services (e.g., for performing reliability predictions, FMECA, and FTA) to the product assurance community.

9.5 Summary

An attempt has been made to cover a lot of ground in a short span. Of necessity, much in the way of details has been excluded. Various meth-

TABLE 9-13 Summary of Product Assurance Analytical Methods

Analytical tool/technique	Description	Concern(s) addressed				
		Requirements Quality assurance	Capability	Conformance	Improvement	Risk
Control charts a. Attributes charts	Provide insight into the stability of a process in which the product's quality characteristics are classed as either conforming or nonconforming to specifications			X	X	
b. Variables charts	Provide insight into the stability of a process in which the product's quality characteristics are either measured or expressed in numbers				X	
Correlation analysis a. Simple correlation	Evaluates cause-and-effect relationship between two variables					
b. Multiple regression	Evaluates cause-and-effect relationship between dependent variable and two or more independent variables				X	
Analysis of variance	Evaluates changes in subject being investigated as affected by several factors				X	
Design of experiments a. Randomized block experiment	Evaluates effects of one factor, randomly applied, on the specimen under study				X	

TABLE 9-13 Summary of Product Assurance Analytical Methods (*Continued*)

Analytical tool/technique	Description	Concern(s) addressed				
		Requirements	Capability	Conformance	Improvement	Risk
b. Latin-squares experiment	Evaluates effects of two factors, randomly applied, on the specimen under study				X	X
c. Factorial design experiments	Evaluates the effects of a number of factors on the specimen under study					
Tests of significance	Evaluate the degree of difference between statistical parameters (e.g., mean, standard deviation, percentages) to determine significance of product or process changes				X	
Process capability analysis	Evaluates the uniformity of the process and its ability to meet specifications		X	X		
Tolerance analysis	Evaluates the product characteristics to permit the establishment of component tolerances	X	X			X
Reliability						
Assessment	Evaluates product reliability performance under test or operational use conditions			X	X	X
Allocation	Provides quantitative reliabilty design goals	X			X	X

155

TABLE 9-13 Summary of Product Assurance Analytical Methods (*Continued*)

Analytical tool/technique	Description	Concern(s) addressed				
		Requirements	Capability	Conformance	Improvement	Risk
Prediction	Provides estimate of product's inherent reliability capability		X		X	X
Failure modes, effects, and criticality analysis	Evaluates the various ways in which a product can fail and the impact on the product's ability to function successfully				X	X
Failure analysis	Evaluates product failures resulting from test or actual operational use to determine failure cause				X	
Worst-case circuit analysis	Evaluates circuit performance under adverse operating conditions		X			
Trade-off analysis	Evaluates effects of changes in reliability capability on other system parameters (e.g., cost, weight, power, volume)	X			X	X
Redundancy analysis	Evaluates the degree of reliability improvement gained by having back-up units available in the event of prime element failure				X	
Sneak circuit analysis	Evaluates the existence of unexpected path or logic flow within a product which, under certain conditions, can initiate an undesired function or inhibit a desired function				X	X

TABLE 9-13 Summary of Product Assurance Analytical Methods (*Continued*)

Analytical tool/technique	Description	Concern(s) addressed				
		Requirements	Capability	Conformance	Improvement	Risk
		Maintainability				
Allocation	Provides quantitative maintainability design goals	X			X	X
Prediction	Provides estimate of product's inherent maintainability capability		X		X	X
Maintenance engineering analysis	Evaluates various maintenance related factors (e.g., maintenance personnel skills, tools and test equipment, facilities, spares, etc.) required to support a product during its operational use				X	
Trade-off analysis	Evaluates effects of changes in maintainability capability on other system parameters (e.g., cost, reliability, etc.)	X			X	X
Assessment	Evaluates product maintainability performance under test or operational use conditions			X		

TABLE 9-13 Summary of Product Assurance Analytical Methods (Continued)

Analytical tool/technique	Description	Concern(s) addressed				
		Requirements	Capability	Conformance	Improvement	Risk
Safety						
Hazards analysis						
a. Preliminary hazard analysis	Identifies gross product hazards during design concept phase				X	X
b. Fault hazard analyses	Evaluates the various ways in which a product can fail and the impact on personnel and product safety				X	X
Quality assurance						
c. Operating hazard analysis	Evaluates product operating functions which are inherently hazardous to personnel or in which personnel error could be hazardous to personnel, the product or both				X	X
Fault tree analysis	Evaluates all various ways in which a product can fail and cause the occurrence of a specific "top" undesired event		X		X	X

odologies that are available to the product assurance practitioner for a given analytical tool have not been discussed. For example, a variety of other methods can also be used for reliability prediction—similar equipment technique, similar complexity technique, prediction by function technique, etc. Discussion of these methods in the detail necessary for their practical use is outside the scope of this book. To cover the methods for all the major disciplines that contribute to product assurance would result in not one book, but rather a volume of books.

Table 9.13 presents a more inclusive summary of the various analytical methods that one can employ. Some of these methods have been discussed in this chapter; others have not. Included in this summary is a very brief description of each method, as well as an indication of the product assurance concerns that it addresses. A quick look at this table serves to confirm the availability of a variety of techniques that can be employed to ensure product integrity.

Manufacturing

Manufacturing quality is everyone's business.
ANON.

10.1 Background

Production, manufacturing, fabrication. These are all names for a general process of assembling parts, materials, and components into some form of functioning item. The manufacturing process itself can vary from an extremely simple stamping operation resulting in a finished penny to a highly complex system of processes requiring many control techniques resulting in, for example, a television set. Whatever the end product and process used, assurance disciplines will enter into the manufacture of it. As a matter of fact, the assurance disciplines started with manufacturing, and the heaviest assurance effort is still concentrated in this part of the overall process of design, manufacture, and use (see Chapter 1). Many organizations still have assurance groups identified only in the manufacturing area. They are generally called quality, quality control, or quality assurance groups. Sometimes they form a part of the manufacturing organization; however the quality organization is preferably maintained separately and reports to upper management.

We will be studying the "quality" of a manufactured product and how a company assures that the requisite quality exists. Quality is the consistent conformance of the end product to the design intent. Design intent is invariably that there exist *no* workmanship or process-induced problems in an end product. No designer of a transistor radio intends *any* of his or her designs to fail because of, say, a high resistance connection; yet some will. No automobile company intends that an axle fracture in use because of a process defect; and yet some axles will. The function of manufacturing product assurance, then, is trying to assure that the design intent is not degraded by workmanship or process-induced defects. However, as in most business enterprises, this function is limited by the resource restrictions imposed by company management. Thus the wide variation in the quality of products can be explained.

The assurance of quality is generally achieved by employing a system that controls received parts/materials and the processes used to assemble the received items into the end product. The control is almost always accomplished by measuring something—the thread size on a screw, the beta of a transistor, the tension of a spring, the rf output of a transmitter, etc. These measurements and tests are used to detect unsatisfactory copies of the product at various stages of fabrication.

The key word is *consistency*. It is preferable to have a consistent process even if it is a somewhat "poor" process—that is, a process that consistently yields 90 percent good results is better than one that varies between 80 percent and 100 percent (even though it may average better than 90 percent) for at least two reasons:

1. There is always the chance that the way we determine "goodness" is not comprehensive, and that the high-variance process is producing unmeasured parameters that fluctuate.

2. We can monitor the first process on a low-sample basis and have more confidence that our sample is representative.

Typical manufacturing functions all have some assurance elements associated with their operation. As shown in Figure 10.1, the elements come from both design and quality assurance discipline areas.

The controlled buying of parts/materials needed for the finished product starts with proper specifications generally developed with heavy design assurance participation. These specifications form part of the design package that defines what is to be manufactured—from the parts through the completed product. The remainder of the drawing package are specifications also—manufacturing specifications, defining the final form assembled parts/materials are to take. The first direct assurance contact with the manufacturing process generally is the inspection of received parts and materials. Following production assembly operations, quality assurance performs in-process inspection, equipment calibration, process control, and other peripheral assurance activities. Often there is some form of testing applied to the finished product prior to its being considered ready for delivery to the customer. This testing can be the visual examination to a drawing requirement or a functional operation to some operational specification. This is usually either

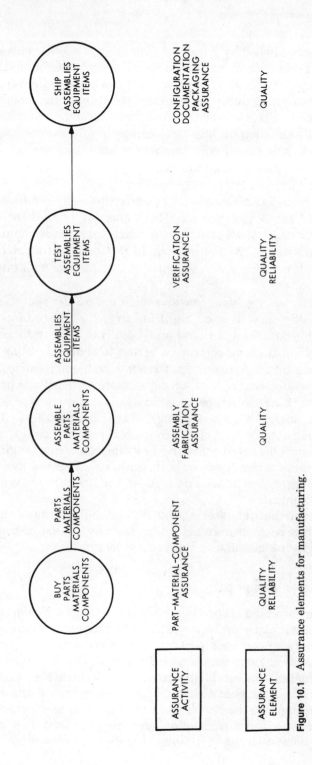

Figure 10.1 Assurance elements for manufacturing.

witnessed or accomplished by a quality function. In a well-controlled manufacturing cycle, a final assurance check is made during packaging and shipping of the finished product. This can be anywhere from a simple check of shipping documentation all the way up to a complete inspection of the packaged product (with some product sampled by removal from the shipping package and complete reinspection).

It should be obvious that quality assurance activities are one with the manufacturing process, being its watchdog, advisor, and conscience. All of these activities are aimed at measuring the consistent conformance of the product—at all stages of manufacture—to specifications developed during the design process. The inspection process includes visual examinations for obvious defects, mechanical measurements for dimensional variation from drawings, and functional testing to operational specifications. The different types of testing are explained in the next chapter.

From all of this testing come considerable data on the operation of the finished product as well as outright failures. Assurance functions should collect this data for additional analysis. This activity should be directed toward an ultimate corrective action to reduce or eliminate the possibility of a future occurrence of the same problem. In some organizations, a separate section, unit, or department is totally dedicated to accomplishing this failure reporting, analysis, and corrective action activity. Most military/government specifications mandate these efforts, but in any industrial process it is an economical part of business operation. Results from this activity are used by manufacturing, engineering, and management personnel in both accomplishing and monitoring progress of corrective actions as well as satisfactory process and manufacturing control.

It is self-evident that the later a problem is found, the more expensive it is to correct it. Assurance activities, therefore, must be concentrated as early in the manufacturing process as practical.

10.2   Assurance Assessment

Beginning with incoming inspection, the aim is to identify parts/materials which might cause problems later in the manufacturing cycle. Significant parameters for each part must be selected for conformance checking, material content must be verified, and fits and tolerances have to be evaluated. The establishment of what to measure at incoming inspection is a somewhat ticklish problem. Assurance can be as minimal as making sure the part is the one ordered or as extensive as measuring every specified parameter on every part received. The important factors to be considered in making this decision include (a) ultimate customer quality requirements, (b) history of the vendor's quality on like parts, (c) criticality of the part use in further assembly, (d) complexity of the part, (e) available inspection budget, and (f) difficulty in troubleshooting/repair of further assembly.

In some instances, where incoming inspection is impractical (e.g., destructive testing or extremely expensive test equipment), the on-site monitoring of a supplier's own assurance measurements is a practical method of inspection.

Statistical sampling techniques are used to measure a representative portion of the received lot to gain confidence that the whole lot conforms to requirements. Statistical techniques and sampling plans have been well developed over time and exist in many textbooks. They also are mandated on many government contracts by MIL specifications such as (a) MIL-Q-9858 "Quality Program Requirements"; (b) MIL-I-45208 "Inspection System Requirements"; (c) MIL-STD-105 "Sampling Procedures and Tables for Inspection by Attributes"; and (d) MIL-HDBK-H53 "Guide for Sampling Inspection." All these techniques depend on suppliers having their own assurance techniques and organizations to ensure that customer specifications are properly interpreted and followed. Assurance money must be spent to check all items to make sure that the few suppliers who deliver nonconforming items are identified. Hence the problem of balancing inspection costs against rework/failure costs must be reckoned with.

In extremely high reliability programs (e.g., a manned space program), where failure and nonconformance is untenable, 100 percent inspection is used more freely. Here, one must recognize that 100 percent inspection does not mean certainty. One hundred percent measurement is actually only about 98 percent effective—due mainly to human and instrument errors. Therefore, a number of 100 percent inspections of different parameters are used to reduce the possibility of unsuitable parts and materials to near zero. In addition to 100 percent functional testing, many types of screening tests have been established for different part types in attempts to weed out parts/materials that are weaker or present a higher failure potential than desired. Examples of this type of screening are:

Integrated circuits

  High-temperature operating burn-in

Semiconductors

  High-temperature bake

  Temperature cycling, thermal shock

  Fine-gross leak

Centrifuge

Shock-mechanical

Relays

Mechanical shock

High-temperature operational cycling

Contact current stressing

Vibration

Resistors

Thermal shock

Power burn-in

Note that many of these screening tests use temperature as a stressing environment. Many studies and experiments have demonstrated the validity of this technique. Also, it is important to understand that some of these screening techniques are utilized by industrial/commercial manufacturers of very complex equipment. For example, computer manufacturers accomplish considerable screening and testing of parts/materials prior to use. It is easy to see why. Consider a computer with 20,000 piece parts. If roughly half of these are transistors, integrated circuits, and diodes, of which 2 percent are defective, a simple calculation shows that approximately 200 functional failures will occur. They will be evident at printed circuit card testing, subassembly testing, final testing of the computer, and—the worse possible place—during customer use. As shown in Figure 10.2, the cost of finding and fixing these failures increases dramatically with progress along the manufacturing-use cycle. These costs accrue from the troubleshooting, repair, and retesting involved, along with the operating and administrative costs associated with maintaining repair parts, special repair techniques, and customer liaison. Accurate knowledge of these costs makes it possible to do an economic trade-off of the amount of parts/material inspection and special testing versus correction cost to determine how far to go in eliminating potential failures.

It is during this activity that initial corrective action identification begins. Consider an electronic device that does not meet its voltage requirements. In order to use it, engineers must be consulted to assure that it will operate dependably in the end product; both the buying function and the supplier must be notified of the nonconformance; and records must be maintained for long-term analysis and possible corrective action. Of course, the best solution for this situation is to send the devices back to the vendor. However, at the present time, many parts are needed by manufacturing personnel because of lead times, schedule slippages, and other delays so it is sometimes impossible to return

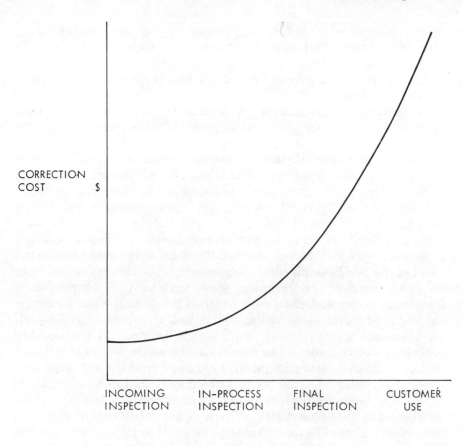

PROGRAM PHASE

Figure 10.2 Correction cost vs. program phase.

the devices. In this example, as in most other instances, the assurance function assesses nonconformance. The final decision as to usability, from a functional viewpoint, comes from design or manufacturing.

In-process inspection is established to assess the conformance of subassemblies (printed circuit cards, power supplies, etc.) to design drawings. Here again, inspection can take the form of visual examination, mechanical measurements, and functional testing. The same economic considerations exist for the assurance of subassemblies as for the parts and materials of which they are composed. One hundred percent inspection is rarely applied across all products; dependence is still on statistical sampling approaches or "roving" inspection—a form of sampling. The statistical sampling techniques used at incoming inspection are entirely appropriate for in-process inspection. In fact, 100 percent in-process testing or visual examination is generally considered to be part of manufacturing processing—under manufacturing direction/control. "Toll gate" inspection (i.e., an assurance function inspecting each and every subassembly prior to continuing the manufacturing process) is only in limited use and is a relatively high-cost assurance approach.

In-process inspection criteria are developed with the same constraints as previously cited for incoming inspection. Since inspection can be accomplished without being directly "in series" with manufacturing processing, it is generally established as a parallel operation having the ability to "shut down" manufacturing. This system works well only with specific quantitative measures established by upper management; e.g., if more than x percent defective are found, corrective action must be taken.

Final inspection is possibly the most important manufacturing assurance point of all. It is the last chance to assure that the product conforms to all of the design-established requirements. For this reason, at least on more complex end products, some form of final inspection is frequently accomplished on each finished item. As the complexity of the end product increases, so does the complexity (hence cost) of quality assurance at the delivery point. Again, as found in subassembly assurance activity, assurance checks can be accomplished in parallel with production measurements. With some end products, e.g., color television sets, samples of the product are placed on 500-h high-voltage life testing to assure consistent quality of delivered sets. There is an excellent business reason for this type of assurance—warranties. No television set manufacturer wants to be faced with an unknown but high number of television sets being returned within the warranty period. The selling price must include a factor for these kinds of repairs, and the manufacturer must be assured that the number of warranty repairs will be within cost requirements. Automobile, washing machine, stereo, musical instrument, and hair dryer manufacturers, to name a few, must also assure themselves of the quality of their delivered products. Another function invariably accomplished by product assurance prior to delivery is making sure that all necessary paperwork is included in the delivery container and that the paperwork matches the product being delivered.

Throughout the complete manufacturing cycle, special fabrication and assembly processes are utilized. These range from simple soldering processes to complex electrochemical plating processes. Any change in the process will result in a change to the resultant product. Again, the quality function has the task of inspection. In this case, it is inspection of the process conformance to specifications, generally established by the engineering branch of manufacturing. As examples, some of the possible types of process control measurements include (a) soldering bath impurity levels, (b) plating bath chemical contents, (c) oven temperatures, (d) furnace gas content, (e) solvent bath contamination levels, and (f) sewage effluent bacteria count. Process variables in the present-day manufacture of sophisticated products can number in the thousands. Their control means the difference between a profitably produced lot of products and a complete loss of both labor and materials.

All of these inspections, evaluations, and assessments provide data. Either attribute data (go–no-go, yes-no, good-bad, etc.) or variables data (5.1 μA, 0.3 V, 10.7 psig, 151.7 tons, 14.2 in, etc.) are the results. These data are needed immediately (to decide whether to continue manufacturing) and should be analyzed to establish longer-term control and trend evaluation.

Many tests exist that cover in exhaustive details the types, application, and value of statistical analysis. In particular, there are a number of texts that deal specifically with the application and presentation of the statistical analysis of assurance data relating to manufacturing (see References). In addition, Chapter 9 covers the basic analytical techniques employed in developing and assessing data to evaluate and monitor assurance trends. However, the most common and useful method of presenting any of this data is the control chart originated by Dr. Walter Shewhart. It is a statistical device generally used to study and control repetitive processes. It allows for the establishment of a standard for a process and exists as a form for recording whether the standard has been attained. As shown in Figure 10.3, the control chart concept is based on the establishment of a desired mean value, an upper control limit, and a lower control limit. These values describe a state of specified statistical control over a time or ordered manufacturing base. For example, if a manufacturer had a gold plating process—primarily for corrosion protection—a control chart could be established to (1) monitor the nominal plating thickness (average), (2) monitor the upper-limit thickness beyond which too much costly gold is used without any added corrosion protection, and (3) monitor a lower-limit plating thickness below which insufficient corrosion protection is achieved. Plating thickness samples could then be taken on a time or production group basis, and the plating thickness measured and plotted on the control chart. If the thickness trend showed a decrease or increase with time or groupings, investigation into the plating bath variables would show why. This information could then be used to establish plating bath chemical percentage specifications to assure that plating thickness remained within specifications.

A specific type of quality assurance control chart is a "fraction-defective control chart (a *p* chart). In this chart, the variable is the fraction found defective in a specific sample versus the samples in manufacturing order. Here, the upper and lower control limits are generally

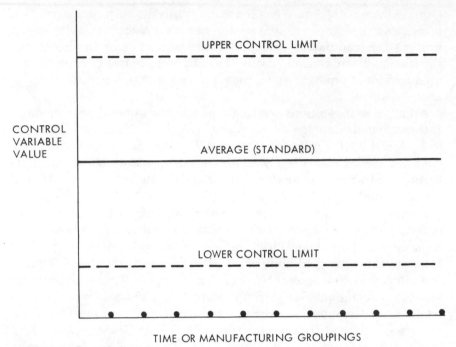

Figure 10.3   Control chart.

statistically established, i.e., within a given number of standard deviations from the desired average value.

In these examples, as well as in control chart theory, when actual values are between the upper and lower control limits, the process or events being measured are said to be "in control." These types of charts provide management an "at a glance" mechanism for monitoring critical processes and parameters of manufactured product. The use of control charts with typical assurance data such as percent defective, yield, failures, etc., serves the basic purpose of reporting the assessment of the product's conformance to design requirements. This is what assurance assessment is all about.

These control limit concepts lead directly into today's computer age. An ever increasing number of manufacturing processes (e.g., plating, cleaning, machining, etc.) are controlled by some type of computer processor. With this type of control and monitoring of processes comes the ability to use statistically determined upper and lower control limits to maintain a given process. Obviously, the same concept can be used to stop the particular manufacturing process if a preset parameter drifts outside of the preestablished limits. With the increase in robotics and automation of hand assembly comes the increased application of the same concept. This whole area of automation has, as its end result, much lower costs for quality and consistent higher quality levels. At present, this technology is in its infancy but is expanding very rapidly. It is driven by an overall drive by U.S. business for lower costs of doing business.

Another extremely critical assurance function throughout the manufacturing process is that of monitoring and controlling nonconforming parts and materials. At every stage of manufacturing, items will be found that do not conform to their individual requirements or the requirements of the end product. Before these items can be used, an objective determination must be made as to whether they should be used at all. During the progress of this decision-making process, the items must be controlled—so that they do not inadvertently find their way into the end product. This control, both of the item and its ultimate disposition, is an assurance function. The records from this process are analyzed for a variety of quality assurance reasons, all of them aimed at initiating corrective action to reduce future nonconformance of the same item for the same reason. The analyzed records provide data for initiating part/material corrective action, design change corrective action, manufacturing process change corrective action, and inspection level corrective action. Quite frequently, the whole process of monitoring, controlling, and taking corrective actions as a result of nonconforming material is formalized into a committee called a material review board. The board usually contains members from design, manufacturing, and purchasing and is usually run by a representative from the quality area. With reject rates at incoming inspection averaging 5 to 15 percent, in-process testing reject rates varying from 5 to 50 percent, and so forth, the importance of understanding and controlling rejected parts/materials cannot be overstated. In any business, the associated costs will form an appreciable portion of the overall manufacturing costs—both in replacement part costs as well as rework. In fact, most government contracts require a formal material review board with a government representative making the final disposition decision.

10.3 Manufacturing Quality

Some of the more common homilies associated with manufacturing quality or product conformance assurance are as follows.

You can't inspect quality into your product.

Quality is everybody's business.

Control the process and you control the product.

Quality is an attitude.

Some of the more common misconceptions relating to manufacturing quality are as follows.

Increasing yield comes from lowering quality.

Quality is controlled by the quality organization.

Quality costs money.

The two diametrically opposed sets of views represented by these two groups of ideas show the problems associated with interpreting the meaning of quality. To some extent, the term's meaning is in the mind of the beholder. Therein lies the problem. Preconceived and developed conceptions of manufacturing quality color and, to an extent, direct the activities of all people associated with manufacturing. The success of any assurance program, therefore, lies in management being committed to the concept of building quality into the product. The extent to which this attitude is positively directed will be a large determinant of the attitude of all manufacturing personnel—down to the lowest skilled assembler. Negatively directed attitudes of manufacturing and management personnel can make producing quality products literally impossible. Over the past 10 to 20 years a major public relations effort has been directed toward educating and convincing all levels of management and manufacturing personnel that their actions are the determinant of manufacturing quality.

Studies which helped to identify the problem of decreasing quality have also indicated that the potential for improving quality exists in manufacturing worker motivation. The early approaches toward worker involvement in manufacturing were established like an advertising campaign. A catchy slogan was invented like "zero defects" (ZD) or "product excellence program" (PEP). Then a public relations communication program would start with top management issuing letters, bulletins, and sometimes taped closed-circuit television programs. Concurrently, sessions would be held with all employees to "sell" them on the concept of involvement in reducing defects, paying more personal attention to what they were doing, and consciously trying to improve performance. "Quality teams," composed of personnel from a variety of organizations, might be established to compete against each other. Measurable goals would be established, results monitored, and awards given to top teams. A considerable variety of these programs emerged in the 1960s and were a definite expression of the need for manufacturing involvement in product integrity.

These approaches certainly were beneficial in increasing the awareness of the whole product assurance field, but the most significant effect is the development of manufacturing *involvement* in quality. There are now many more manufacturing personnel who have begun to understand the necessity for having involvement in the quality discipline—both economic and philosophical.

As a result of foreign economic competition *and* better quality of foreign imports, some very innovative production approaches have been attempted. The most significant is the "production team" that combines most of the elements of manufacturing supervision and quality control. These teams are allowed to make decisions about working hours, production quotas, quality levels, etc. They invariably are responsible for a much larger than normal portion of a complete production process—sometimes all of it. These teams compete with one another against a number of criteria (cost, defects, etc.), and, of course, personal involvement is significant. This concept also is being used by companies in a number of foreign countries. Japanese manufacturers have a "quality circle" concept which is like the older "zero defect" programs in the United States. In all of these, there is the attempt to increase the motivation of the "hands on" workers to produce items with workmanship pride, a solid indication of the recognition that manufacturing product integrity is the responsibility of the manufacturing organization, not the quality organization.

11

Testing

I shall prove it—as clear as a whistle.
                                    JOHN BYRON

11.1 Introduction

The most nearly unimpeachable source of product integrity information is testing. Personally, we have a tendency to regard "proven by test" claims with the same suspicion generally accorded "proven by analysis" claims. Among the great masses, however, "seeing is believing." Witness the television commercials in which the watch still runs after serving, with the propeller, to drive a motorboat to racing victory; the pen still writes after being used as a can opener; the leak-sealing antifreeze (and the laws of fluid mechanics) preserves the integrity of its original container; the upset-stomach remedy provides a protective (or at least opaque) coating on the inside of a beaker; the battery simultaneously starts a fleet of automobiles; one spray deodorant can be used to stick cotton balls to the inner elbow, while another one can't...ad nauseam. The advertising agencies have perceived that the consumer would like to "try before buy," even if only vicariously, via the boob tube. This natural and reasonable desire is the motivation for most testing. We applaud both the desire and honest response to it, even by advertising agencies. (We have come to praise testing, not to bury it.). As far as we know, the watch, pen, and battery demonstrations cited were valid and to the point, if not comprehensive. Even the deodorant stickiness test is pertinent, if one considers stickiness and body odor to be equally odious. The upset-stomach remedy demonstration was not a test, but merely a visual aid. It was graphic illustration of the nature of a claim—not substantiation. The attempt to imply substantiation was sheer flimflam, as was the "demonstration" of the leak-sealing property of the antifreeze. In this case, the real hero was the partial vacuum formed in the top part of the can.

Such flimflam is not confined to the consumer market, nor practiced exclusively by advertising people. We know of a case in which production reliability testing was required for a product to be used by the military. The way this works is: Every so often a random sample of the product was allowed to operate for a certain time—failures were recorded and repaired as they occurred. If more than the predetermined maximum number of failures occurred, the product failed the test. The entire production lot from which the sample was taken was deemed unacceptable, and production was stopped, not to resume until the cause of substandard reliability was found and fixed. In this particular case, the procedure was to select a slightly larger sample, operate its members in a stress situation (burn-in) for a time, and then submit the appropriate number of survivors to the reliability test. This has two obvious effects: (1) the chance of flunking the reliability test was reduced, because most "infant mortality" defects were avoided, and (2) the items submitted to the reliability test were not representative of the original population from which they were selected and which they were *supposed* to represent. An even more ingenious circumnavigation of the reliability demonstration problem: the first hours of a reliability test were designated as "burn-in," so that failures during this period, if they occurred, would not count in the accept/reject decision. To be "fair," however, neither would these initial so many hours count toward reliability test completion. For items which experienced no failures during this initial period, however, this initial time was counted as valid demonstration time. Among statistical mathematicians, this is known, technically, as "having one's cake and eating it, too." Now, neither of these cases involved covert subversion of good product assurance principles. Everything was open, documented, and approved by the proper authorities. It was all quite legal, it just wasn't *right*.

What is testing? According to our Funk & Wagnalls: TEST tr. verb (a) to try by subjecting to some experiment or trial, or by examination and comparison; (b) to subject to conditions that reveal the true character of, in relation to some particular quality. So, basically, testing consists of providing stimuli to the test subject and noting reactions.

11.2 Kinds of Tests

There are three primary reasons for testing: to seek knowledge, to verify a hypothesis, and to classify. In addition, there are testlike processes designed to eliminate inherent weaknesses, purify, make better. And, of course, motivation for testing comes in various combinations of these factors. A test may be designed to intentionally destroy the test subject, either because of the nature of the subject or the objective of the test. It should be obvious that destructive tests relate to "class questions" about the test subject. If conclusions drawn from such a test cannot be applied to the test subject's brothers as well as the test subject itself, then there is no point to the test. Nondestructive tests, on the other hand, might direct themselves to specific test subjects—with no extrapolation of results to others of their kind. Thus, there are really just two *kinds* of tests, namely:

1. *Class tests*.   Class-question testing seeks information or makes determinations related to item types, families, or classes. As a consequence, it is important that the actual test subjects chosen be truly *representative* of the population about which information is sought or for which some determination is to be made. Class testing may be either destructive or nondestructive, and applied to hardware or software.

2. *Item-directed tests*.   Item testing makes determinations related to the specific item under test only—and must therefore be nondestructive. Preconditioning processes are included in this category, since they quite obviously have no effect on eliminating weaknesses from items not included in the processing (except as a by-product). These kinds of tests are solely hardware oriented.

For a worm's-eye view of the world of testing, see the "test apple" of Figure 11.1. It is arranged with like characteristics grouped in segments with the most basic test characteristics at the core and growing more specific toward the outside. Design-related testing occupies the left hemisphere and manufacturing testing the right hemisphere. Those tests most closely related to the assurance of product integrity overlap both, and occupy the lower hemisphere.

Ideally, from the point of view of accuracy and confidence of results, all testing would be conducted on each and every item of interest. Test results would be deterministic rather than probabilistic, and the test would, of course, be nondestructive. On the other hand, from an economic and practical point of view, it would be ideal if the results of testing a single representative item could be confidently applied to the entire parent population of that item. Both of these situations, and the many compromise positions in between, actually exist in the scientific/industrial environment.

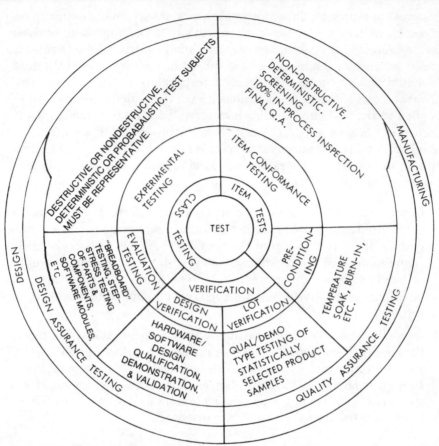

Figure 11.1   The test apple.

It should be noted that software-related testing is only briefly mentioned here and is covered in more detail in Chapter 13, Software Quality Assurance.

11.2.1   Class testing

Within the category of class testing, there are two major subclassifications. One is *experimental testing*. This type of testing simply seeks knowledge, though usually with some general objective in mind. Experimental testing fits exactly the definition: to subject to conditions that reveal the true character of, in relation to some particular quality. It embraces product assurance *evaluation testing*, which is generally directed at determining the suitability of various parts/components and software modules for inclusion in the product design. Evaluation testing is also useful for determining reasonable values for part/component characteristic specification limits, methods, and procedures for incoming/in-process verification testing, design derating policy, and stress levels for preconditioning processes. Evaluation testing can also be applied to the product itself—e.g., indefinite life testing. The second major type of testing is *verification testing*. It differs from experimental testing in that the objective *is not* to determine the true nature of the subject in relation to a particular quality. Verification testing is less ambitious, being content to determine merely whether the test subject meets certain preestablished criteria, e.g., that the software will execute all the proper commands in the correct sequence. It may be that a given verification test *does* establish the true nature of the characteristics under examination, but in verification testing, this is a fortunate by-product, not an objective. There is a clear distinction between design verification and as-manufactured product or lot verification. In design verification, the *representativeness* of the test subject needs careful qualitative substantiation, for the results of the test apply to *all* copies of the product (i.e., to the product design). In lot verification, the representativeness of the sample is established statistically, and the test results apply only to a specific subset of the total product population.

11.2.2 Item-directed testing

Item-directed testing, like class testing, embraces two major subclassifications. One is *item conformance testing* (such as in-process factory tests). It is applied to each item of the population of interest and is go/no-go (like verification testing). However, it is rather cursory in that only major performance parameters are ordinarily examined. Like verification testing, item conformance testing requires predetermination of accept/reject criteria. Parts screening is a form of item conformance testing wherein defective or substandard parts are eliminated from incorporation into the product, via 100 percent inspection of the incoming parts. Parts screening is often used in conjunction with stress preconditioning—to assure the consistent reliability of a particular part type or family, in the end product. The second type of item-directed testing is *preconditioning* or *burn-in*, in which all items of interest are subjected to controlled stress conditions so that weaknesses can be detected and discarded or repaired. The survivors are then known to be "tried and true." Preconditioning can be applied to any level of assembly, from the piece-part to the entire product. Quite often, a necessary prerequisite to establishing the stress conditions for preconditioning is a well-thought-out experimental or *evaluation test*. The stresses must be of the proper sort and severity to precipitate detectable failure of inherent weaknesses, but at the same time they *must not* cause permanent damage. Another troublesome consideration of preconditioning is *time*. How long does it take, under the stress conditions imposed, to rid an item of all, or most, of its latent defects?

11.3   Test Planning

Well, now that we have completed this brief categorization of test varieties (according to basic test objectives), let us consider some of the more important cares and cautions surrounding test planning. Few endeavors need more careful planning than a test, and documentation (remember Chapter Seven). If a test isn't documented, it isn't done, and that applies to the test plan, the test procedure, and the test results. It doesn't necessarily have to be fancy, but it has to be done. The need for proper documentation of test results is obvious. It is easy then, to conceive of situations in which the absence of companion plans and procedures could render the results less credible, or less useful.

11.3.1   General considerations

There are a few basics to be considered in the planning of any test. Are we contemplating a test to find out something we would like or need to know (seek knowledge—experimental testing); to develop evidence that will substantiate some preconceived notion (verify a hypothesis—verification test); or to separate the apples from the oranges, or the sheep from the goats (classification screening, preconditioning, or item conformance testing)? At this point, alternatives should be sought and considered: Does there exist any credible history which will supply my needs directly? Can the previous, or concurrent, work of others be analytically extrapolated to satisfy my needs? What confidence (either qualitatively, or statistically quantitatively, as the case may be) do I require in the results?

Having established that a test is necessary, or at least desirable, we must next survey the resources available for performing the desired test—time, machinery, and personnel. Timing—when and for how long—may be important to the project schedule. It is obviously futile to schedule the test of a particular item (either hardware or software) before a truly representative copy has been developed and fabricated. Necessary machinery includes the equipment for providing input stimuli (signal generators, temperature/humidity/altitude chambers, load press, vibration table, etc.), diagnostics exercisers in the case of computer-driven equipment, the instrumentation for noting test conditions and test subject reactions (meters, gauges, etc.), and the actual test subject. Personnel requirements are qualitative as well as quantitative, i.e., personnel skilled in the detailed aspects of test planning, equipment operation, etc., are required.

11.3.2   Environmental testing

The most common problems encountered in environmental testing (examining or verifying performance in the presence of artificially created environmental conditions) are associated with either: (1) the necessity for interfering with operating environments that are unrelated to the test objectives (e.g., the unwanted effects of temperature chamber walls on the propagation characteristics of an antenna) or (2) the natural tendency of measuring devices to affect the quantity being measured (e.g., the problem of inserting a sensor into a dynamic equilibrium loop without affecting the parameters we wish to measure). We have an example which illustrates both sorts of problems.

The problem   Measure the friction torque presented by a bearing system, as a function of temperature.

The solution   Cut holes in the bearing housing to allow a detectable strain (elastic displacement, twisting) in the presence of the anticipated range of stress (torque—force transmitted to the housing from the driven shaft, because of bearing friction) as shown in Figure 11.2. Place strain gauges on the weakened section of housing to detect the strain, which will be proportional to the friction torque of the bearing system.

The subproblem   The strain gauges will also be affected by the varying temperature, and give different readings at different temperatures for the same torque.

The solution   Run correlation tests to develop known temperature correction factors for the strain gauges.

The pitfalls   One might ask, "Why not simply measure the variation in energy required to drive the shaft?" This is a good question, and worthy of pursuit. However, we feel it would be difficult to separate the friction torque effect of the bearing system under test from the effects of losses in the driving motor itself. We would be faced with such difficulties as passing a thermally nonconducive shaft through the temperature chamber wall (another bearing surface). In short, we run the risk of masking the effect we wish to measure to the extent that it becomes a threshold phenomenon. We would then enter the realm of what Dr. Irving Langmuir called, "the science of things that aren't so"[1]. This is the kingdom in which Finagle's laws are operative (First you draw the curve, then you take data which plots on the curve. The proper correction factor is equal to the desired results divided by the observed results,..., etc.)

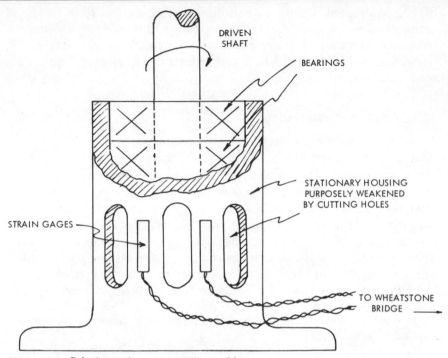

DRIVEN
SHAFT

BEARINGS

STATIONARY HOUSING
PURPOSELY WEAKENED
BY CUTTING HOLES

STRAIN GAGES

TO WHEATSTONE
BRIDGE

Figure 11.2   Solution to instrumentation problem.

On the other hand, recognize that there are many situations in which a properly planned, executed, and documented test is the best available solution. These situations are characterized by:

1. The need to gain information, verify a hypothesis, or classify the test subject

2. The absence of sufficiently credible analytical techniques and/or applicable historical information

3. The availability of the resources necessary to achieve the objectives

The *need* for a test is ordinarily judged on a technical basis. However, a politically motivated need can be just as real. If the boss wants some information that he or she doesn't really need, why we have to supply it, don't we? Just be cautioned that "the science of things that aren't so" was born of just such needs—needs generated by the "discover or die," "publish or perish" syndrome in our institutions of science and academia (such pressures are legitimate enough, it's just that they are sometimes applied to the wrong people, at the wrong time, for the wrong reasons). Likewise, when the *credibility* of alternatives to testing is examined, we must ask, "Credible to whom?" It may not be enough that *we* believe that a theoretical analysis, or extrapolation of history, will satisfy the need. The boss, the customer, the scientific/industrial community, or public opinion may disagree. In any case, if you attempt to conduct a test without the necessary *resources*, you are almost certainly doomed to walk Finagle's nether regions, for a time, and obey his laws.

Don't let paranoia about test flim-flammery cloud the fact that there is no substitute for a properly conceived, planned, executed, and documented test. There are those who contend that "the only *real* evidence of product 'goodness' is a history of user satisfaction," particularly in regard to product integrity characteristics. While there is agreement that the *best* evidence of product integrity is continued user satisfaction, attempts to use such statements to deny the efficiency of testing are largely erroneous. Finally, conclusions based on valid test results have two important advantages:

1. *Timeliness.*   Decisions can be made in time to avoid a lot of grief.
2. *Utility.*   Test data lends itself readily to the determination of cause-and-effect relationships and is very often useful for purposes not originally anticipated. This is possible because of the controlled conditions and the comprehensive recordkeeping that is part of testing.

11.3.3 Reliability growth testing

In the area of product reliability, an important concern is to ensure the maturity of the design as it goes into production and later into the field. An immature design will result in residual design flaws being detected in manufacturing, rather than during development, with costly engineering changes being the penalty. As a result, reliability growth testing has emerged as a vehicle for finding and eliminating design weaknesses at a timely point in the product life cycle. This type of testing is being used more and more as a precursor to, and sometimes as a substitute for, the kinds of statistical verification testing described in the next section.

Reliability growth testing should follow environmental testing. The growth test samples must have previously completed environmental testing, and all necessary fixes identified through the latter testing incorporated in these samples.

Several approaches to reliability growth testing can be pursued.

1. Test the samples, find a deficiency, develop a fix for that deficiency, incorporate it into the samples, and then retest. Presumably, if deficiencies are effectively corrected, they will not resurface in the next interval of time and the reliability of the product (as reflected by the test samples) will steadily increase.

2. Test, find a problem, work a fix, but do not incorporate it until some opportune time in the test program (but before its conclusion). Since problems are allowed to persist until their fixes are incorporated, if these fixes are effective, the retest will show a significant jump in product reliability.

3. Use a hybrid of the previous two. Some fixes are incorporated as deficiencies are found, while others are delayed and incorporated at some later convenient time. In this case, product reliability improvement will be characterized by periods of steady growth in some intervals of time and leaps in other periods.

Whatever the approach used then, reliability growth is achieved by the systematic and permanent removal of failure mechanisms stimulated by stress-oriented testing. Design defects are aggravated or forced to surface, and corrective action in the form of effective redesign is implemented. The key is not just to find these deficiencies, but to fix them so that they don't happen again. The proof is found in a plot of reliability growth versus time (as well as in the nonappearance of previously discovered failure mechanisms).

One of the more commonly applied reliability growth plotting tools is based on the work of J. T. Duane while at the General Electric Company (in the early 1960s). An example of the Duane plot is shown as Figure 11.3. Product reliability (in terms of MTBF) at the starting point (i.e., after testing for approximately one multiple of MTBF) is set at somewhere between approximately 0.1 and 0.3 of the predicted MTBF as estimated through analysis of the paper design (see Chapter 9, Analytics). Reliability growth (in terms of MTBF) is plotted cumulatively (total unit time divided by total failures) over time on log-log paper. The instantaneous MTBF is determined by multiplying the cumulative MTBF by the reciprocal of 1 minus the slope (α).

11.4   Statistical Verification Testing

For a variety of reasons, the product characteristics of reliability, maintainability, and quality cannot be verified, with absolute certainty, by use of testing. However, given unlimited resources, such characteristics can be verified to any desired degree of statistical confidence. The

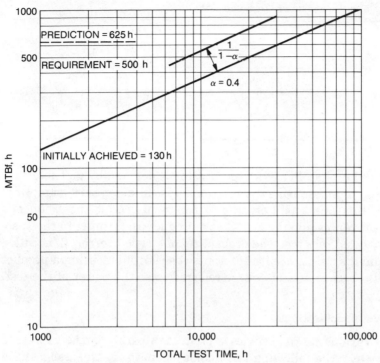

Figure 11.3   Duane plot example.

basis of a decision test is the theory of testing hypotheses. In product assurance testing, the hypothesis under test is that the submitted product conforms to some specified product integrity requirement. An alternative hypothesis is that the submitted product does not conform to the requirement. Rejection of the first hypothesis is essentially equivalent to acceptance of the alternative hypothesis. Since samples rather than the whole population are inspected, the possibility of incorrect inferences due to sampling fluctuations must be considered. If H_0 represents the hypothesis that the product conforms to the requirements and H_1 represents the alternative hypothesis that the product does not conform to the requirements, either of the following two incorrect inferences may be drawn:

1. *Type I error:* H_0 may be rejected when it is true.

2. *Type II error:* H_0 may be accepted when H_1 is true.

The probabilities of making these errors are dependent on the sample size and decision criteria. If the probability of a type I error is denoted

TABLE 11.1   Test Decisions vs. True Situation: Risks

Test Decision	True Situation	
	H_0 True	H_1 True
Accept H_0	Correct-decision probability $= 1 - \alpha$	Type-II-Error probability $= \beta$
Reject H_0	Type-I-error probability $= \alpha$	Correct-decision probability $= 1 - \beta$

by α and the probability of a type II error is denoted by β, Table 11.1 can be used to summarize the relationships. In acceptance sampling, the probability of a type I error, α, is commonly called the producer's risk since it represents the risk that a product conforming to the specification will be rejected. The probability of a type II error, β, is called the consumer's risk, since it represents the risk of accepting a product that should be rejected. $(1 - \beta)$ is referred to as the power of the test.

11.4.1   Test specifications

Since the various characteristics of product integrity can be expressed in many ways, the following terminology is used for this generalized discussion of the determination of objectives of statistical product integrity testing:

1. Acceptable integrity level (AIL) is the level of product integrity, measured by an appropriate parameter, considered to be acceptable, i.e., $H_0 = $ AIL.
2. Unacceptable integrity level (UIL) is the level of product integrity considered to be unacceptable and representing the alternative hypothesis, i.e., $H_1 = $ UIL.

The AIL will be the value that has a high probability of acceptance, and the UIL the value that has a low probability of acceptance. It is important that the values of the AIL and UIL be consistent with product use requirements.

No credible means has yet been devised to verify product integrity (or even systems effectiveness, dependability, or availability) by a single test. Product integrity must be verified via a composite of reliability, maintainability, safety, longevity, and demonstration tests, coupled with evidence of effective quality control (including proper statistical quality verification testing). The point made here, however, is that the basic principles of statistical testing are identical whether we are

TABLE 11.2   Commonly Used Test Specifications

Product Integrity Characteristic	AIL (acceptable integrity level)	UIL (unacceptable integrity level)	Usual Measurement Parameters
Reliability	θ_0 (acceptable reliability level)	θ_1 (unacceptable reliability level)	number of failures per unit of operating time
Maintainability	AML (acceptable maintainability level)	UML (unacceptable maintainability level)	duration of maintenance actions; mean, median percent of sample performed within a specified interval
Quality	AQL (acceptable quality level)	LTPD (lot tolerance percent defective)	number of defective items out of random sample

concerned with reliability, quality, hardware, software, or the effects of sunspots. Table 11.2 shows the relationships of the more important product integrity characteristics to AIL and UIL.

Ordinarily, reliability (expressed as MTBF) is specified in terms of both θ_0 and θ_1, either directly or indirectly. For instance, the acceptable *reliability* level θ_0, and the discrimination ratio θ_0/θ_1; or the *unacceptable reliability* level θ_1, and the discrimination ratio may be fixed. The indirect specification accounts for the term, "probability ratio sequential test" (PRST). Maintainability is usually specified in terms of acceptable maintainability level (AML) only, but frequently with constraints on the distribution of times-to-repair (i.e., the mean corrective maintenance time shall be less than or equal to 30 min; and 90 percent of all maintenance actions shall be less than 1 h). The unacceptable maintenance level (UML) is either at the discretion of the experimenter, or it is indirectly specified by imposition of some particular maintainability demonstration test plan (e.g., one of the test plans of MIL-STD-471). Quality control inspections and tests are normally constrained by either specified AQL or LTPD, but not both.

A mastery of trade jargon is essential to the product assurance engineer in the major institutional (Department of Defense, Army, Navy, Air Force, Federal Aviation Administration, Department of Transportation, etc.) supplier environment. An understanding of the realities of the powers and limitations of statistical testing is required, in *any* case.

The α and β risks represent the specified decision errors associated with good and bad incoming products, respectively. Since lower risks require more testing, a balance between the level of test effort and the cost of a wrong decision is sought. Although the level of test effort is predictable, the cost of wrong decisions is not. For this reason, only test costs are normally considered in specifying α and β. Sometimes several tests are performed sequentially. The first is a design-proof type. If this type of test is scheduled early enough to allow design changes, α and β may be relatively high (.20 or more), since the main objective is to assure that the product is not completely unacceptable. For the formal demonstration test, i.e., a test that judges conformance to the specified requirement, the α and β risks are generally lower (on the order of .10).

Test requirements are sometimes specified in terms of confidence levels. Such a specification is subject to serious misinterpretation, if it is not precise. For example, assume that a specification states that a sample shall be tested to determine with 90 percent confidence that the equipment conforms to the requirement of a 0.5 h mean time to perform corrective maintenance ($\bar{M}_{ct}$). Two reasonable test procedures would be:

1. Compute the 90 percent lower confidence limit L. Since there is a 90 percent confidence that the true mean time to repair is greater than L, if $L \leq .5$, reject the equipment; otherwise, accept it.

2. Compute the 90 percent upper confidence limit, U. Since there is a 90 percent confidence that the true mean time to repair is less than U, if $U \leq .5$, accept the equipment; otherwise, reject it.

Test 1 is equivalent to one in which the producer's risk is 10 percent at a true mean of 0.5 h. Test 2 is equivalent to one in which the consumer's risk is 10 percent at a true mean of 0.5 h. The difference between the two tests is that the former requires that equipment with a true mean corrective maintenance time of 0.5 h be accepted 90 percent of the time while the latter requires acceptance only 10 percent of the time if the true equipment $\bar{M}_{ct}$ is 0.5 h. If a test specification is to be made in the form of a confidence interval, it is imperative that it be made clear whether the specified characteristic represents an acceptable or unacceptable level.

The operating characteristic curve. By specifying two of the three quantities, n (sample size), α (producer's risk), and β (consumer's risk), the accept/reject criteria of the verification test is uniquely determined for a given family of tests. It is then possible to construct the operating characteristic (OC) curve of the test plan. This curve shows the probability of acceptance of all possible incoming product integrity levels. Two points on the OC curve are already determined—the α and β points with their corresponding integrity levels, which are given by H_0 (AIL) and H_1 (UIL), respectively. For example, assume that the product integ-

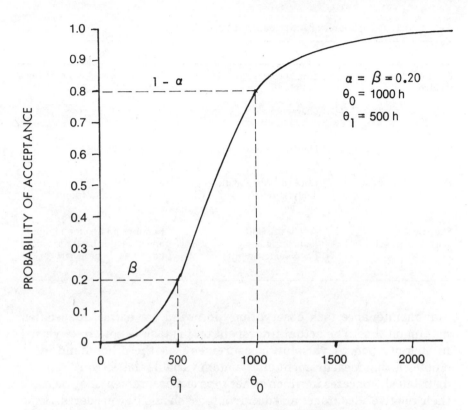

TRUE MEAN TIME BETWEEN FAILURES, h

Figure 11.4   Typical operating characteristic curve.

rity requirement under test is in terms of mean time between failures (MTBF) and that the AIL is θ_0 = 1000 h MTBF and the UIL is θ_1 = 500 h MTBF. The α risk is 0.20, and the β risk is 0.20. The general shape of the OC curve will then be as shown in Fig. 11.4. The probability of acceptance may be viewed as the long-run proportion of equipment or lots that will be accepted. For example, if the OC curve shows that equipment having a MTBF of 800 h will be accepted with a probability of .65, then, if all incoming equipments are tested, 65 percent of all incoming equipment with a true MTBF of 800 h will be accepted.

Attributes and variables.   The data type is usually categorized in terms of attributes (classification data) or variables (measurement data). For example, in maintainability testing, the usual attribute type test is one in which a success/failure determination is made on each sample observation according to some preestablished criterion. Thus, if the maintainability requirement is related to maximum duration of repair times, a corresponding attribute measurement would be that a partic-

TABLE 11.3   Comparison of Attributes and Variables Tests

Factor	Attributes Test	Variables Test
Type of Information yielded	Number or percent of sample that meets some specified characteristic	Observed distribution of some quantitative output
Sample size requirements	Higher than variables test for corresponding plan	Lower than for attributes test for corresponding plan
Ease of application	Data recording and analysis relatively simple	More clerical and analysis costs than for attribute plans
Statistical considerations	Applies to both parametric and nonparametric tests	Requires an assumption on the underlying distribution unless large sample properties are assumed

ular maintenance task observation did or did not exceed a specified maximum time. The actual time spent on the task is not directly used in this decision. A variable measurement, on the other hand, does employ actual measurement of a random variable that is continuously distributed. For cases in which either type of measurement may be used, their relative advantages and disadvantages should be considered. Some of these are summarized by Table 11.3.

A parametric test is one in which the underlying probability distribution of the random variable is assumed to take a specific form. In reliability testing, for instance, an exponential failure probability distribution is usually assumed and the validity of the test depends on the real existence of this distribution. Nonparametric or distribution-free tests are those in which no assumptions about the underlying probability distribution are made. Nonparametric tests do not deal with magnitude, but with attribute characteristics such as frequency and ordinal position.

Sampling.   Single, multiple, and sequential sampling plans can generally be devised so that each affords the same degree of protection. For convenience in this discussion, we shall assume an attributes test wherein a sampled item, or observation, is held to be a success or a failure according to whether the measured value is less than or greater than some specified value. Table 11.4 compares various sampling plans.

TABLE 11.4 Comparison of Various Sampling Schemes

Characteristic	Single Sampling	Multiple Sampling	Sequential Sampling
Sample size	Known	Average can be computed for various incoming quality levels—generally less than single	Average can be computed for various incoming quality levels. Generally less than single and multiple
Decision choices	Accept or reject	Accept, reject, or take another sample until final sample is selected	Accept, reject or test another item
Predetermined characteristics	Two of the three quantities n, α, and β	Same as single	Fix α and β; n is a random variable.
Personnel training	Requires least training	More trained people required than for single	Requires most training.
Ease of administration	Easiest—scheduling can be fairly precise; test-cost estimates can be made	More difficult than single since the exact number of tests is unknown. Only average test costs can be estimated	Most difficult in terms of testing scheduling, and overall administration —most time consuming.
Miscellaneous	Best used for testing situations where ease of administration is most important and cost of testing is relatively unimportant	Has psychological advantage in that supplier is given a "second chance" by taking further samples if first-sample results indicate a marginal lot	Most efficient test in terms of required sample size. May require approximately 50% of sample size of single-sampling plans. Best to use when test costs are most important

In single sampling, one sample of n items is tested (or n observations are made). Accept or reject decisions are made by comparing the number of observed unacceptable events with a predetermined acceptance number c. In multiple sampling, more than one sample may be necessary to reach a decision, but the maximum number of samples is known. An example is a double sample plan:

n_1 (first sample size) = 20 c_1 (accept number, first sample) = 3
n_2 (second sample size) = 40 c_2 (accept number, both samples) = 7

A first sample of 20 items is taken. If 3 or fewer unacceptable items are found, an accept decision is made. If 8 or more unacceptable items are found, a reject decision is made. If 4 to 7 unacceptable items are found, a second sample of 40 items is taken, and an accept decision is reached if the total number of unacceptable items is 7 or fewer.

Sequential sampling is an extension of multiple sampling, in that decisions to accept, reject, or sample further can be made after each individual observation (or, possibly, groups of observations). For a standard sequential plan, no maximum number of sample items or observations is specified, although the probability of very large sample sizes is ordinarily quite small. The decision criteria for a sequential sampling plan can be presented graphically. Figure 11.5 illustrates a sequential test where the number of unacceptable outcomes is the decision statistic. As the sampling progresses, the number of unacceptable outcomes is plotted against the number of trials. Testing continues until the plotted step function crosses one of the two decision lines. Since the step function may remain in the continue-test region for a long period, especially for borderline lots, truncation or stopping rules can be determined so that the effect on the α and β errors is negligible.

Concluding comments on statistical testing.   Generally, parametric tests are more efficient than nonparametric tests since, for a given amount of testing, more precise estimates (or smaller probabilities of incorrect decisions) will result. The limitations on the types of statistics that can be determined constitute a disadvantage of nonparametric tests. For example, nonparametric tests of central tendency apply only to the median, while the mean may be of more interest, in a particular situation.

It is emphasized that an incorrect assumption of the underlying probability distribution in a parametric test can lead to an OC curve quite different from that planned.

11.4.2   Bayesian tests

A bayesian test can be defined as one which uses prior information in the decision criteria concerning the random variable of interest. So far, classical-type tests have been discussed. The classical test involves decision criteria based on prescribed probabilities of acceptance for specified product integrity levels. In the bayesian test, the results are combined with the prior information (prior test or analytical results) to yield a revised (bayesian) estimate of the actual distribution or parameter, and decisions are made according to the desirability of this estimate. The classical tests defined here are well documented, have been shown to provide the necessary protection against rejecting good or accepting bad product, and thus have been accepted as being a

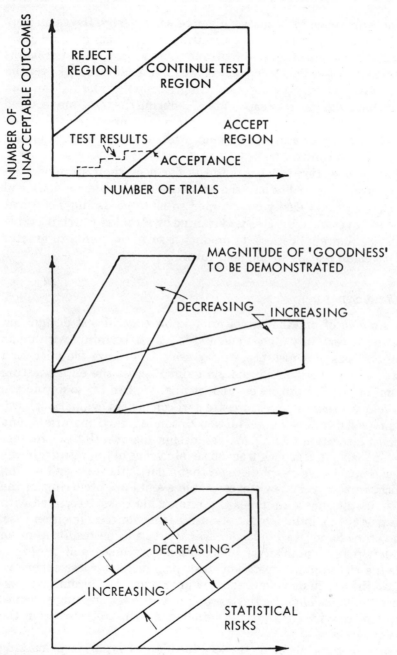

Figure 11.5 Some basic aspects of sequential testing.

reasonable approach toward assuring product integrity. Bayesian tests are still not widely implemented because of skepticism in some quarters, and their application will continue to undergo a trial and learning process. However, they do possess two distinct advantages over the classical test: (1) they provide for using available information, and therefore have the inherent capability of reducing the test time required before reaching a decision; (2) they can provide assurance on the distribution of outgoing or accepted product, while classical tests generally provide no such control.

The major objection to bayesian tests has been their strong dependence on a prior distribution—the existence of which some deny and others claim cannot easily be obtained so as to be useful. The great interest in bayesian statistics, as evidenced by recent research in statistical theory and applications to product assurance, points to greater use of bayesian testing in the near future.

11.5 Test Synergism

The basic *idea* of making use of prior knowledge in test design and evaluation is certainly a good one and is just as certainly within our present abilities. For instance, we can seek to optimize the allocation of test resources versus information gained across the entire testing program for a given project or product line. We can try to devote the proper relative amount of resources to part/component evaluation, hardware and software design evaluation, incoming parts/materials, and in-process inspection testing. We can design the overall test program to be responsive to feedback from our evaluations of the results. In the midst of a project, we might discover that a particular test isn't buying us much, in that very few unacceptable events are occurring in the test. We might choose to increase the tolerable risks associated with this test, or even eliminate the test—decreasing our costs in either case. On the other hand, we might discover the test to be resulting in an inordinately high number of unacceptable outcomes and choose to reduce the risks on an "upstream" test (presumably, a less expensive test than the one in question), thereby improving the "incoming integrity level" of items entering the test in question. Thus, we might reduce overall test cost by reducing the number of items recycled from the test in question.

When you think about it, the key to this test synergism—making the value of the total test program equal to more than the sum of the individual tests—is an information system that will allow a singleentity to evaluate all test results. This implies organizational relationships and data retrieval methods which are subjects unto themselves.

These considerations notwithstanding, we must keep in mind that one of the objectives of a product assurance test program is to force product integrity improvements as early in the product design/development/production cycle as possible.

Note

1. I. Langmuir, "Pathological Science," transcribed and edited by R. N. Hall, General Electric Report No. 68-C-035, April 1968.

12

Improvements

*A common mistake is to think of failure as
the enemy of success. Failure is a teacher—a
harsh one, but the best.*
                              Thomas J. Watson, Sr.

12.1 Role of Assurance

Preceding chapters have described both technical and management techniques for achieving product integrity. These techniques are intended to minimize the need for "downstream" product improvements—to make the development of the requisite product reliability, quality, etc., a matter of routine. Perfection is seldom achieved, particularly with regard to "new" products. In the course of the creative process, the product undergoes constant "improvement." The typical hardware designer is dedicated to design improvement without end, the software engineer is always striving for performance enhancements, the manufacturing engineer is constantly seeking less expensive fabrication methods, the quality engineer works toward "zero defects," and so on, into the night. These aims are counterbalanced by management cost and schedule objectives. The proper role of the assurance sciences in guiding and contributing to this process of product evolution is examined here. Because of its importance, the notion of improvements is discussed again separately as it relates to software (see Section 13.10).

The mission of product assurance is to see to it that the product possesses the required degree of integrity and no more. This is the overall aim, as biased by the proper management goals of product assurance. However, individual practitioners of the assurance sciences are no more immune to provincialism than are members of the other engineering disciplines. Quality control inspectors tend to develop personal quality standards above requirements, safety engineers take a morbid interest in one-chance-in-a-million hazards, and reliability engineers want life test data without end. Further, witness the Japanese and their unwavering dedication to continued product improvements.

Now, all this is as it should be—given proper management. The designer will not allow the reliability engineer to eliminate a desirable functional feature to achieve greater reliability; the quality assurance engineer will not sit still while the manufacturing engineer's new process compromises consistent quality, and so forth.

The proper role of the assurance sciences in product improvement is played by an effectively managed group of eager, aggressive, and technically competent product assurance specialists. Their aim is to seek improvements in product integrity as required to reach a predetermined level of durable utility—with a license to seek further improvements only in those cases resulting in increased *value* to the producer. For example, replacing a discrete part in an electronic circuit with a monolithic microcircuit (even though the original circuit is entirely satisfactory) may result in increased reliability and decreased cost, weight, volume, and power consumption.

The most significant set of tools the assurance specialist has for identifying the need for improvements is an efficient "intelligence" system. A rather wide variety of these systems exist. Their common objective is to detect and correct product deficiencies, thereby making necessary improvements. These systems are applied throughout the product's life cycle, making use of most of the assurance tools described in preceding chapters. These deficiency detection and corrective action schemes are the chief initiators of necessary or desirable improvements in product integrity.

12.2   Corrective Action Systems

Assurance means reduction of risk through the resolution of foreseeable problems and difficulties. A farsighted search for these problems and their effective resolution is a goal of effective assurance. The system for detecting, recording, resolving, and reporting problems is the means through which visible assurance is obtained. Most of the assurance tasks discussed previously have as a component the revelation of problem areas that could hinder further development. For instance, analyses turn up nonconforming designs; tests reveal deficiencies; design and program review result in action items; and failure reporting results in action to correct the failure condition.

Corrective action systems come in all sizes and shapes and span the product life cycle. Although different in structure and time phasing, these systems are common in purpose—to anticipate, detect, analyze, and correct problems which adversely affect product integrity. This activity is a closed loop feedback process in which problems are continually brought to the surface and resolved.

As the product proceeds through design, fabrication, test, and field phases, various problem recognition techniques are brought into play. In the early program phases, design review provides the means for problem anticipation. Later, assessments show the way to these problems.

The methods of problem solution run the gamut from design changes during the design phase to field modification and new software releases during the user-operational phase. The most cost-effective solutions are those which are initiated early in the program; the earlier, the better. For example, a different value resistor can be easily introduced during design through a simple drawing change. The same change to the resistor value during production would require hardware modifications which could have significant schedule and cost implications. During user operation, the resistor value change requires either field modification kits or, worse, return of all products. Thus, we have the general situation in which the cost effectiveness of improvements decreases as programs mature. Unfortunately, the relative ease of problem recognition is inversely related to the cost effectiveness of problem solution. The reason for this is twofold: (1) The applicable problem-recognition techniques accumulate as the product moves through its life cycle, enabling us to have more tools at our disposal during later stages. (2) This is compounded by the relatively uncertain nature of potential problem anticipation as opposed to problem detection by hard data. And this is particularly true in the case of software.

A key point in the closed-loop feedback process which characterizes effective corrective action systems is *follow-up*. It is not sufficient to simply recognize problems and initiate corrective action. Any action taken must be followed up to ensure that it is effective. In short, corrective action systems must lead to real improvements in product integrity. These techniques are used to obtain improvements across all program phases.

12.3 Design Improvements

An effective design review program is perhaps the producer's most valuable product assurance tool. The design review process is the first line of defense against product mediocrity. It presents the first and most cost-effective chance to detect and correct hardware and software product deficiencies. Subsequent to the design concept, design reviews are used to discover deficiencies and initiate improvements. There are two categories of product improvements—mandatory changes which must be made to assure the requisite product integrity and "improvements-of-opportunity." Design review, then, is oriented to the detection and correction of instances wherein the design does not satisfy requirements, with a weather eye out for opportunities to increase value.

The design review process includes many activities in addition to (or indeed, in the absence of) formal inquisitions. Design review ranges from a program of scheduled formal meetings with the attendant breadboard testing and reporting, parts/materials application studies, trade-off studies, simulation techniques, etc.—to the individual designer critiquing his/her own design, or talking it over with his/her coffee buddy. The aim, and quite often the result, is the same. However, it is noteworthy that in the case of the individual designer performing his/her own design review, any change is most likely to be concerned with improved performance. Further, it is possible that this improved performance may have to be purchased at the expense of money, time, reliability, or something else of value. It is desirable, therefore, that the designer have some training in the assurance sciences, or that his/her coffee buddy be a product assurance specialist. The engineer who is trained in the assurance sciences has an appreciation for such subtleties as:

1. The producibility of a design, i.e., the facility for being economically fabricated with consistency of quality

2. The reliability of a design, i.e., the possession of the ability to endure its environment uninterrupted by malfunction during active use and to withstand the rigors of time in addition to the vagaries of chance

3. The human fallibility and physical limitations of average equipment operators

4. The physical dangers of high potential energy differentials to operators and maintenance personnel

While it is unrealistic to ask an individual hardware design engineer to take note of *all* these factors, the proper design review process must review all pertinent design factors against the need for improvement. The same holds true for software. The attributes of software quality (see Section 13.2 Elements) are extremely diverse and require special attention by software quality assurance personnel.

Mandatory design change requirements are most often determined to be in the areas of performance, reliability, operability, maintainability, safety, and software quality. Mandatory performance improvements are left in the capable hands of the responsible design engineer—except to say that such improvements should be carefully examined to assure that overall product integrity does not suffer unnecessarily as a result.

The need for design reliability improvement is mostly a result of design oversight in the required durable quality of product elements— ± 5 percent resistors do not necessarily stay within ± 5 percent of their initial value for an extended period. Occasionally a part is simply misapplied (e.g., a natural rubber gasket sealing the gasoline line).

Operability is sometimes compromised by disregard for a precise, documented definition of product malfunction. A radar operator is confused if she suspects that the behavior of a target might be the result of an equipment malfunction. The automobile driver wonders, when his oil pressure "idiot light" glows while idling at an intersection, whether his oil pressure is adequate.

The maintenance person is bemused by malfunction symptoms with conflicting trouble-source indications, or situations that require the whole product to be dismantled in order to repair a high-failure-frequency item (e.g., the need to take the whole washing machine apart in order to replace a V-belt).

Leaving safety hazards that could have been eliminated in the design phase of a product can be disastrous. Witness the Ralph Nader publicity of recent years and the death of the Corvair.

Soft, or recoverable, errors, if allowed to persist and accumulate in software, result in a significant degree of customer dissatisfaction.

Note that the key to design improvement in the assurance science area is *anticipation* of problems. Such anticipation is achieved only by analysis and test. The test phase is effective only as a result of careful analysis. Successful analytical anticipation and prevention of problems is singularly unspectacular; failure alone is spectacular.

Obviously, the proper time to catch a design mistake is before it is realized as a product—and certainly before it gets a public airing. However, regardless of the time or method of detection, something must be done to bring the product up to par. In the case of design deficiencies, this is a matter of analytically devising the required design change, empirically altering the design. This process isn't necessarily easy, but it is relatively inexpensive, so long as a factory hasn't been tooled up to manufacture and test the original design or, perish the thought, if the original version hasn't already been built, especially in quantity; or the software product hasn't been released to the field. In such difficult cases, it may be worthwhile for the producer to circle the problem. For instance, imagine a case where the defective item has been produced and a large quantity is in use. If the problem is one of reliability (not involving personal safety), the producer might well elect to put a corrective design change or software patch or workaround into effect for current and future production hardware models or software versions and simply endure any warranty claims and customer dissatisfaction with earlier hardware models or software releases. In general, however, design inadequacy must be corrected by design change, accompanied by whatever retooling, reprocurement, field retrofit, or software patch is necessary to assure the required degree of integrity for all copies of the product. The advantages of early detection and correction of design deficiencies are not limited to the producer's own house. If subcontractors are depended on for the design of components, it behooves the producer to monitor this activity closely to assure that they have effective design review programs. Incidentally, if a subcontractor attempts to "circle the problem," the producer should have provided for effective contractual means to keep them to the straight and narrow, within reason.

It should be obvious by now that design deficiencies are perhaps the most serious lapses in product integrity and that their correction grows increasingly costly and embarrassing as the product moves through its evolution and application. Early detection and effective correction of such deficiencies is essential to product integrity. Correction can only be assured by assiduous follow-up.

12.4   Production Modifications

In the production phase, during which fabrication and test take place, a variety of assurance systems are employed to improve product integrity. Whereas in the design stage, the emphasis was on problem anticipation, the shift is now to problem detection. Nevertheless, the goal is the same—product improvements. Since the factors that can be detrimental to product integrity are diverse, improvements must be aimed in many directions. Therefore, production modifications are defined as any improvements, initiated during the production phase, which enhance product integrity; e.g., design changes, process changes, etc.

The various assurance systems employed during the production phase have several common activities: (1) reporting of product deficiencies and related events, (2) analysis of reported data, and (3) initiation of corrective action. As the product proceeds through the production phase these activities are exercised. Improvement need is identified through the analysis of data generated from assurance audits and controls. Corrective action is then taken to satisfy the identified need for modifications. Thus, assurance audits and controls provide the means for acquiring data for the analytical process, whose output is problem identification leading to corrective action. Since production modifications can be both good and bad, follow-up is necessary to verify their effectiveness.

During the procurement stage, a set of assurance requirements are prepared by the producer and imposed on any suppliers. As a program proceeds through the design phase, design activities of the suppliers are reviewed and improvement opportunities developed as previously described. Finally, in the production phase, the supplier's ability to manufacture the item is monitored closely. Process specifications are reviewed, manufacturing facilities visited, process controls examined, and tests witnessed. The producer can monitor these and other activities to ensure that the supplier is meeting requirements. Monitoring these activities can lead to problem identification before shipment of parts and materials are made to the producer. The supplier can be made to correct deficiencies through the provisions set forth in the contract.

Prior to shipment, the producer may perform source inspections at the supplier's facility. Nonconforming items can be rejected at this point, and the supplier made to take corrective action. More typically, purchased parts and materials undergo an incoming inspection at the producer's facility. Rejected items can be reviewed by a materials review board to establish their disposition—either accepted "as is" where minor discrepancies are involved, or returned to the supplier for rework and/or replacement where major discrepancies are prevalent. The producer can maintain inspection records and discrepant material reports on each supplier and review these records to identify suppliers of inferior quality components. Other problem detection tools include supplier performance indices, quality control charts, and lot plots. Consistently inferior quality can signal the need for major process changes by the supplier, or in extreme cases, the removal of the supplier from the producer's approved source list. The latter approach, however, is not without hazard and is usually considered only as a last resort—the sole source problem. In this case, the production schedule must allow for the evaluation and selection of an alternate source.

During fabrication, process controls are imposed to ensure product uniformity. Among these, quality controls are used during in-process and final assembly to provide the data needed to detect sources of excessive deficiencies. The data have many forms—from basic product yield data providing "number accepted out of total inspected," to detailed inspection data listing "number of defects by type." Another good tool for the identification of production problems is the statistical quality control (SQC) chart which is described in Chapters 9 and 10. The SQC chart can be employed at various product fabrication stages to track the rate of defects or rejects. The chart utilizes probability "warning" limits to provide a timely signal of an out-of-control situation. Regardless of the technique used, once the sources of product rejects and contributing defects have been identified, appropriate prevention approaches can be brought to bear. For example, if poor soldering is the major cause of excessive product rejects, then defect prevention could take the form of a soldering training class aimed at improving workers' soldering skills. Follow-up on the effectiveness of the prevention approaches taken can be performed through continual review of inspection records and yield data to establish that a downward trend in rejects/defects is actually being realized.

In the test phase, the producer has additional opportunities to check product integrity and, if necessary, initiate improvements. The cost of undertaking these improvements is now much higher since the product is further "downstream." As the product (including its lower assemblies or components) goes through various tests, pertinent information is accumulated. As before, test yield data are recorded, SQC charts are maintained, and so on. In addition, a failure report is filled out to document all the circumstances surrounding the occurrence of any anomalies. The failure report is reviewed to determine if the failure cause is apparent and whether failures of a similar type have been experienced in the past. In some cases, the "paper" analysis will have to be supplemented by physical analysis of the failed part itself to zero in on the failure cause. Determination of the cause of failure is important since the course of corrective action must be aimed at this cause. It is not enough to know the number of failures of a given part type. In order to assign corrective action, the "how" and "why" of a failure must also be established. Once the cause of failure has been determined, the responsibility for corrective action can be assigned. For example, if a number of transistors are found to be "burned out" in a given piece of equipment, failure investigation could reveal the failures to be caused by a test technician rather than by poor circuit design. Thus, an effective failure reporting, analysis, and corrective action system can be used during the test phase to prevent failure recurrence.

In addition to their primary role as valuable verification techniques, the demonstration and qualification tests described in the previous chapter also serve an important secondary purpose—they provide a basis for initiating improvement in product integrity. The failure data from these tests, when properly analyzed, will lead to problem identification and act as a trigger for incorporation of product improvements. Appropriate assurance parameters (e.g., mean life and repair rates) can be quantitatively determined and compared with established numerical requirements. Noncompliance may require that production modifications be made to meet these requirements. Life test data can be evaluated to ensure that expected product life is consistent with the guarantee statements. For example, the producer might take a dim view of offering a 5-year guarantee on a light bulb if 90 percent of those bulbs consistently failed to provide the guaranteed lifetime under life test; obviously, product improvements (or guarantee rewording) are in order. Finally, qualitative assessments can be made of test results to detect other product deficiencies, such as potential safety hazards, which must be corrected through improvements.

Thus far, various means for problem detection have been described. More to the point, however, is the matter of problem resolution. How does the producer make the necessary production modifications happen? First, the *type* of modifications required to improve product integrity must be determined. Then, the *mechanism* with which to execute these modifications must be selected. Finally, the incorporation of product improvement must be followed up to ensure that it is effective, for only then is the problem detection-evaluation-correction loop closed.

Production modification types	Executive mechanism
Design changes	Change engineering description of product
Manufacturing changes	Change manufacturing/build instructions
Use changes	Change user instructions

In summary, during the production phase, a variety of assurance activities take place that will lead to product improvements. Proper execution of these activities, culminated by the analysis and use of the resulting data, can provide the means for problem identification. Investigations into these problems can yield recommendations for their solution. With adequate follow-up, the effectiveness of this corrective action can be ensured.

12.5   Retrofits

Producers learn of their mistakes through some form of use and service problem reports. This reporting can take on many forms—from an all encompassing documentation system to the salesperson discussing the product with the user. In any event, field reporting is the way the producer finds out how well the product is operating, how well customers like it, how much it is costing for maintenance (when the producer is responsible), and whether improvements are needed. It was possible in the dim past for producers to maintain ignorance of the results of using their product. Now however, with liability suits, increasing government controls, and competitive warranty pressure, the necessity for good field reporting should be easily understood. Some of the forms field reporting can take are service/software problem/repair reports; failure, salesperson, and marketing reports; and user complaints. These can be either verbal or written. Whatever the information source, it is mandatory that an identified person or organization (e.g., software hotline group) be responsible for the collection, investigation, and coordination of the data and resulting actions. Future sales and reputation can be lost when a user is unable to obtain a personalized review in answer to a problem (whether real or imagined). The assurance function must provide an ideal centralization point for activities associated with usage problems and their resolutions.

An ideal field reporting system for a producer should include:

1. A good communication link for problems, ideas, and recommendations

2. A central point of contact at the producer for data grouping, analysis, and corrective action

3. A summary reporting system to management which shows major trends (cost and performance)

4. A feedback system to the user and field personnel

Many of the identified usage problems will be resolved by discussion, changes in use and maintenance techniques, and actual repair. Occasionally, however, the bane of all producers arises—a field retrofit. The necessity for sending personnel, a kit of parts, and additional equipment can be very troublesome. A field retrofit might even require the recall of a "lot" of products, e.g., automobiles. Retrofit programs usually occur for reasons of safety, gross malfunction, or severe user dissatisfaction. The best way to avoid them is to apply effective assurance techniques during design, development, and production. However, even with the application of these techniques, problems do get to the user. Two examples follow.

Example 1   A very large number (thousands) of power supplies were used in each of a number of phased array radar systems installed in various parts of the world. A stringent product assurance program existed to preclude unsatisfactory reliability, safety, etc. The power supply subcontractor was required to design, develop, and test the power supply to assure sufficient reliability and make it nonrepairable (throw away). One year from the day the first radar system went into continuous operation, power supplies began failing at an alarming rate. Subsequent investigation and analysis involving international travel revealed a main filter capacitor failure. The problem was internal anode lead contamination which caused chemical decomposition when power was applied. Electrochemical reaction required about a year of operation before the lead was completely decomposed. The ultimate result was that the government, the prime contractor, the subcontractor, and the capacitor vendor shared the cost of a very expensive field retrofit program which included modification of the power supply to make it repairable.

Example 2   A major manufacturer of bed sheets recently received a customer complaint stating that the first time the sheets were washed, they were torn in the washing machine. The manufacturer's investigation revealed that one of the chemicals used in the processing was much too concentrated during the manufacture of one batch of material and had severely weakened the fibers. The result was the removal of the batch from retail stores, replacement of sheets for those customers who complained, and irreparable loss of reputation with those customers who had not.

In addition to responding to problems, real improvement possibilities can be identified in a product by user reporting systems. (Note: This aspect of improvements is raised as it applies to software in Section 13.10 Improvements). The same communication channels through which field problems flow should be used to supply ideas and recommendations for improvements from field service personnel and users. Better mouse traps, automobiles, and electronic systems can result from this process. They can then be sold for a higher price or used for a competitive advantage to gain more sales.

12.6 Economic Considerations

It is almost axiomatic that improvements will cost the producer additional instant money. What may not be so obvious is that *some* instant money spent can actually result in long-term money saved. Assurance programs designed to spend money judiciously during early project phases do save money in the long run.

In most projects, money spent on design reviews, process control development, safety analyses, reliability studies, software field testing, and other techniques is a good business investment. These assurance techniques will minimize the possibility of customer dissatisfaction, field retrofit, and recall programs, legal liability entanglements, and other equally expensive results of product improbity.

However, it is possible to go overboard; product improvements can be sought to their ultimate. Thus, proper management attention to schedule and cost constraints must be exerted to reduce this possibility. The constructive conflict derived from these two somewhat divergent goals is generally beneficial to a project. This conflict forces the development of improvements within specified requirements.

Making the product "too good" can result in reduced sales of field services, spares, and general usage support. Planned obsolescence, although not generally held in very high regard, suffers considerably from this approach, to the detriment of future profits.

One final note of warning—the number of contracts (particularly in government business) that result in law suits by the producer to obtain funds for "cost growths" (overruns) is increasing. Quite frequently, these law suits arise because of improvements included without benefit of contract change. Although settlements are usually obtained, they help neither the producer nor the government in the long run. Similarly, in the commercial area, improvements without contract change or customer desire can be disastrous.

The watchword for assurance participation in product improvement is *value*. Improvements must be aimed at maximizing profits. Assurance techniques provide the means to this end.

Chapter 13

Software Quality Assurance

Software is the result of an intellectual process; it doesn't fail!

ANON.

13.1 Introduction

As noted earlier in Chapter 1, History, the use of computers has created a new set of challenges related to the assurance of software integrity. This chapter provides some insight into the way the assurance sciences are dealing with these challenges.

The matter of assuring software integrity goes under the popular banner of "software quality assurance"—thus, the title of this chapter. However, in the context within which we have used "quality assurance" in this book, this title can be misleading. As discussed earlier in Chapter 2, Elements, quality assurance has been associated with ensuring that the designed-in integrity of a product is not degraded by the manufacturing process and that consistency is achieved. In the case of software, however, consistent "manufacture" of the product is not the primary assurance issue (except in the case of microcode, or software embedded in hardware, such as in read-only memories). Instead, the quality assurance efforts are aimed at the design, development, and test of the software. In essence, software quality assurance becomes more closely aligned with the design assurance function, rather than with manufacturing assurance and will be so treated in this book. To see this better from a product life cycle viewpoint, Figure 13.1 shows the phases of the software life cycle interleaved with the traditional phases of the hardware/system life cycle.

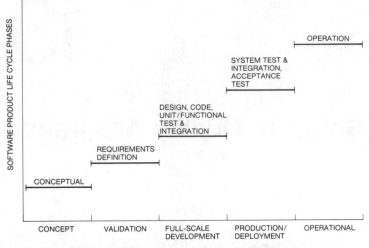

Figure 13.1   Hardware/system product life cycle phases.

In terms of structure, this chapter draws upon many of the same subheadings found in other chapters: elements, organization, planning, controls, documentation, design, analytics, testing, and improvements. However, whereas these chapters previously treated the software assurance issue superficially, this chapter provides a more focused, topical discussion. Nevertheless, consistent with the rest of the book, this chapter is restricted to an overview of software quality assurance and how the function ties together the various tasks and activities to assure the integrity of the software product.

13.2   Elements

The attributes of software quality are best portrayed in diagrammatical form, as in Figure 13.2, and related to the three major phases of the software product life span: operation, maintenance, transition.

In order to comprehend better the role of software quality assurance, it is important first to understand what each of these attributes mean. Thus, each attribute is defined below.

- *Correctness.*   The extent to which a piece of software satisfies its specification and fulfills the system objectives of the user

- *Reliability.*   The extent to which software can be expected to perform its intended function with required accuracy

- *Efficiency.*   The amount of computing resources and code required by the software to perform a function

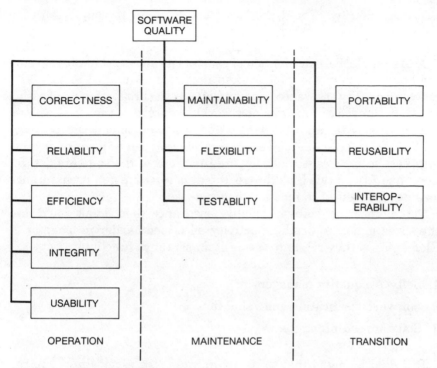

Figure 13.2   Software quality attributes.

- *Integrity.*   The extent to which access to the software or its data by unauthorized persons can be controlled

- *Usability.*   The degree of effort required to learn, operate, prepare input, and interpret the output of the software

- *Maintainability.*   The degree of effort required to locate and fix an error in an operational piece of software

- *Flexibility.*   The degree of effort required to modify an operational piece of software

- *Testability.*   The degree of effort required to test software and ensure that it performs its intended function

- *Portability.*   The degree of effort required to transfer software from one hardware configuration and/or software system environment to another

- *Reusability.*   The extent to which software can be employed in other applications related to the packaging and scope of the functions that the software performs

• *Interoperability*. The degree of effort required to couple software systems with each other

A contemplative examination of each of the above attributes reveals the broad scope of the quality assurance role as applied to software. Software quality assurance as a function has the broad mission of assuring that the software works and works accurately; that it is secure; that it is easy to use, fix, change, and test; and that it can be used, reused, and interfaced. Another observation that can be made by reviewing these attributes is that they are functions of the software design. Consequently, in this book the treatment of software quality assurance as a design assurance function makes sense.

The elements of software quality assurance then break down into the three major segments of activity described earlier in Chapter 2, Elements, as they relate to the design assurance function:

1. Software quality management

2. Software verification and validation

3. Software maintenance

It should be noted that software maintenance is not limited to the development cycle; it nearly always extends to the released software as modifications (updates) are introduced. Consequently, software quality assurance fits nicely into the commonality of purpose and approach of the assurance sciences by serving to prevent, detect, and correct deficiencies related to the above attributes across the product life cycle to achieve total integrity of the software product.

13.3 Organization

From an organizational viewpoint, there are typically three approaches which are employed to position the software quality assurance function:

1. As part of a manufacturing quality assurance function

2. As part of the software development function

3. As part of a dedicated assurance function

Each of these organizational approaches is examined in terms of its effectiveness in fulfilling the software quality assurance mission.

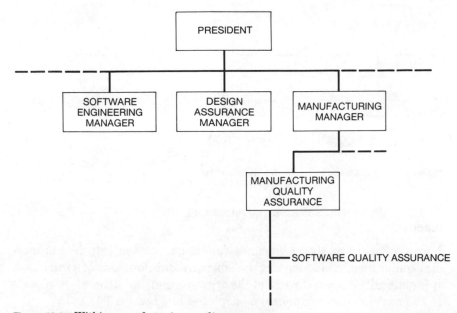

Figure 13.3   Within manufacturing quality assurance.

13.3.1 Within manufacturing quality assurance

Of the three approaches, manufacturing quality assurance (see Figure 13.3) is clearly the worst place to position the software quality assurance function in the overall organization. This positioning of software quality assurance generally is the sign of an immature organization which does not understand the role or responsibility of software quality assurance. The need for such a function is probably perceived, but a sufficient understanding of where to place it is lacking. However, since a quality assurance function (associated with manufacturing) exists, the mentality is to simply "bury" it within a group with a like name.

The effectiveness of a software quality assurance function embedded as a subset of the manufacturing quality assurance organization is extremely low. It typically serves as little more than an accounting function to track what's going on but has little influence. The people in the function are generally weak in software development expertise and experience, and thus are rarely respected; software quality professionals are difficult to attract to this type of function because it is tied to hardware manufacturing. As a result, a "we-them" syndrome surfaces between the software developers and the software quality assurance people, so effectiveness suffers.

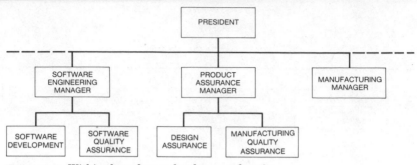

Figure 13.4   Within the software development function.

13.3.2 Within the software development function

Many software quality assurance functions, particularly in commercial companies, reside within the software development organization. It is generally a separate arm of the organization, but ultimately reports to the development manager, as depicted in Figure 13.4.

This type of organizational approach is analogous to the "degenerate approach" described in Chapter 4, Organization. The software quality assurance function is part of the same organization whose work is being critiqued. It is the awkward situation of the "wolf guarding the hencoop." Yet, the effectiveness is not as bad as it would appear at first glance. Since the people involved are typically of similar software backgrounds and experience, there appears to be a mutual respect between the developers and the quality assurance people. Thus, the interaction is one of peers, rather than adversaries. Nevertheless, when both functions report to the same top manager some inappropriate trade-offs and compromises in software quality sometimes results.

13.3.3 Within a dedicated assurance organization

For the same reasons brought out earlier in Chapter 4, Organization, a more ideal way to structure the software quality assurance function organizationally is within a dedicated assurance organization. As a reminder, the latter type of organizational approach offers the characteristics, and advantages, of

- Getting top management attention

- Maintaining a core of assurance professionals in the face of economic pressures

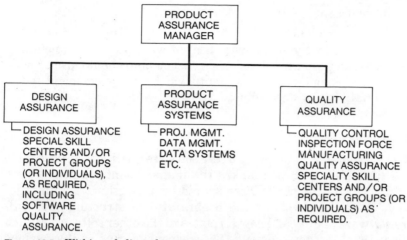

Figure 13.5   Within a dedicated assurance organization.

- Maintaining a close working relationship with other functional groups (particularly the assurance groups)

- On a par structurally staying with the engineering (development) organization

An expanded version of the ideal product assurance organization proposed in Figure 4.7 (see Chapter 4), modified to specifically reflect the software quality assurance function within design assurance, is shown in Figure 13.5. This organizational approach offers a high degree of uncompromised, unbiased, focused effort on software quality on a more efficient basis.

13.4 Planning

The old adage, "Fail to plan; plan to fail,"applies to the assurance disciplines just as readily as it does to others. The importance of product assurance planning was highlighted earlier in Chapter 5, Planning. In addition, a product assurance planning model is provided in Appendix A to give the reader general guidance in structuring and tailoring a plan to a specific product. Nevertheless, because of the emerging nature of software quality assurance and the substantial contribution software makes to the performance capabilities of today's products, further separate treatment is offered here for software quality assurance planning.

13.4.1 Introduction

Surely, the plan should contain information regarding its purpose and scope. It should clearly define what the software product is, including all its pieces, and what it will be used for. The plan should also provide a complete list of referenced documents which will be pertinent to the software quality assurance effort.

13.4.2 Management

The lack of consistent positioning of the software quality assurance function within an organization makes the management segment a very important part of the plan. There should be a description, with organization chart, which shows the organizational structure that will assure software quality and how it relates to the other software-related functions, such as development. The responsibilities of the various participating groups should be succinctly defined so that there is no confusion about who has prime and supporting roles for each of the tasks to be performed. There should also be a definitive schedule of tasks and activities to be implemented as part of the software quality assurance program, complete with major milestones and deliverables.

13.4.3 Task descriptions

The plan should provide task descriptions in sufficient detail to allow for tracking and control of the software quality assurance effort. Each description should provide sufficient insight into dependencies and interdependencies, inputs required, task outputs (deliverables), procedural information as to how tasks will be performed, and reference to other plans, where appropriate. As a minimum, task descriptions should be provided for the following:

- Establishment of software specification requirements
- Performance of software reviews and audits, such as specification requirements reviews; design and code reviews; and functional, physical, and in-process audits
- Implementation of software verification and validation testing
- Implementation of a configuration management program
- Implementation of a problem-reporting, analysis, and corrective action process
- Implementation of control functions for code, suppliers, and media

Embedded within each of the above task descriptions should be further information regarding software documentation standards (e.g., for logic structure and coding), practices, and special tools, techniques, and methodologies (e.g., simulation and software performance measurement) to be employed.

Each of these task areas is discussed further in later sections of this chapter.

13.4.4 Reporting

The last major segment of the plan should spell out the reporting requirements of the software quality assurance program in terms of frequency, content, and format.

13.5 Controls

In terms of controls, the same kinds of technical, cost, and schedule approaches described in Chapter 6, Controls, apply to software quality assurance. The matrix in Table 13.1 illustrates this point very well as the reader can see the applicability of these approaches to either hardware, software, or "system" (encompassing both hardware and software) products. One additional approach, which is unique to software quality, has been added here to those described previously in Chapter 6, namely evaluating the cost of software maintenance.

13.5.1 Evaluating the cost of software maintenance

The use of computers has brought about another kind of special cost, one related to software maintenance. It is definitely a cost to be reckoned with. Many of the large computer companies have software hotlines to help customers through knotty problems in running the software. Large staffs, backed up by software support specialists, are put in place to listen to the customers' problems, help in their diagnosis, and offer solutions so that they can continue to do useful work. These software problems are generally documented in the form of software problem reports, (SPRs), and the count of SPRs and status toward the problem's closure is diligently tracked. Sometimes the solutions are simple and related to user errors or poor documentation. Sometimes the people on the hotline can offer temporary workarounds or "patches." And sometimes, if the software problem is critical and/or widespread, the more costly solution of another software release is undertaken. In any event, software maintenance costs embrace the cost of

TABLE 13.1   Applicability of Control Approaches

Approach	Hardware	Software	System
Technical			
Requirements response control	X	X	X
Design control	X	X	X
Parts/materials control	X		
Design change and configuration control	X	X	X
Subcontractor and supplier control	X	X	X
Manufacturing control	X		X
Corrective action control	X	X	X
Cost and Schedule Control			
Evaluating the relative cost of assurance	X	X	X
Utilizing cost forecasts	X	X	X
Utilizing cost-to-complete estimates	X	X	X
Evaluating the cost of defects	X		X
Evaluating the cost of test failures	X	X	X
Evaluating the cost of maintenance service contracts	X	X	X
Evaluating the cost of warranties	X	X	X
Evaluating the cost of software maintenance[*]		X	
Utilizing milestone charts	X	X	X
Utilizing PERT	X	X	X
Utilizing C/SPS	X	X	X

[*]Only this approach is discussed here; see Chapter 6, Controls, for details on all other approaches.

staffing the hotline and making any necessary changes in the software. If the software has considerable errors and is not easily maintainable, software maintenance costs will soar well above what was expected. These costs can be monitored and corrective measures taken to reduce them to a tolerable level.

13.6   Design

As noted at the beginning of this chapter, the major challenge for the software quality assurance function occurs during the design and development of the software. It is for this reason that software quality assurance is treated as a design assurance function in this book.

The attributes of software quality (refer back to Figure 13.2) are broad in scope and are concerned with more than just error-free software. The time to influence the software product so that it embodies these attributes is as early as possible during the design/development phase. Toward this end, therefore, the software quality assurance function participates in the software development process to ensure that the proper focus and influence is being placed on incorporating these attributes into the software. The remainder of this section describes the software quality assurance activities that typically are performed in the pursuit of this objective.

13.6.1 Establishment of software requirements

Product requirements. Various surveys have shown that one of the leading causes of software nonquality is inadequate or incomplete specifications. Obviously, inadequate or incomplete requirements can cause difficulties in programming, in verifying/validating compliance, and most importantly, in meeting user expectations. The consequences are also obvious in that software patches or revisions need to be generated, software maintenance costs escalate, and customer dissatisfaction abounds. And yet, this area is still given insufficient attention.

The key to establishing a "good" specification is to start with the product requirements. It is important that the latter requirements be thorough and specific, and truly reflect the market needs. The best way to satisfy market needs is to take the users' perspective of the software product. In the best case, actual characteristics of the product will match the users' expectations, needs, and requirements. As a minimum, the mismatch will be very small. The real challenge, however, is to narrow the gap for a wide range of users, perhaps having different expectations, needs, and requirements. Therefore, inputs from various groups having close ties to users, such as marketing and customer services, will be highly valued.

The product requirements should be related to the software quality attributes (e.g., maintainability). Furthermore, they should be measureable and verifiable. In the case of a new release of a current software product, the existing product can be evaluated in quantitative terms to see how it is perceived by users. Then the focus can be directed toward removing the deficiencies which cause the existing product to be perceived as lacking in quality, and enhancing the new release accordingly. For a new product, measurable/verifiable goals should be developed on a priority basis and aimed at satisfying the broad range of expected users. In either case, a high degree of emphasis and commitment should be placed on the task of establishing concise product requirements.

Specification requirements. The product requirements then are translated into specification requirements (or goals). Again these latter requirements should be both measureable and verifiable. There is some merit to stating the justification for a particular requirement to aid in zeroing in on why it is there in the first place. In addition, it is sometimes useful to document the nongoals in the specification as well to eliminate any later confusion. The specification will go through a number of reviews and changes during the definition phase and then be firmed up before the development phase begins.

The specification should contain details around the environment in which the software product will be employed, covering user profiles (e.g., prior computer experience of the users), use environment (e.g., secretarial work station), usage patterns (e.g., extent/duration of usage), and work flow patterns (e.g., work tasks to which the design is related). The hardware that the software will be used with, as well as other software that it will interface with, should also be specified.

The capabilities of the software product should be completely described in the specification. The functional, operational, and usage aspects of the product should be complete. This section of the specification is extremely important to the development of clear user documentation. It should contain a top-level description of software functionality and descriptions of all discrete software elements (e.g., device handlers) covering functions performed, inputs/outputs, processing, user interaction, etc. Functional performance requirements should be defined for inputs, processing, and outputs. Inputs should be expressed as to their source, format, quantity, timing, range, scaling, and method of receipt. The processing requirements should be described in text or mathematical form and include functional parameters as well as any geometric diagrams. Both internal and external outputs should be defined to cover their method of transmission, timing, format, destination, meaning, scaling, and range. Furthermore, strong consideration should also be given here to user interfaces, such as provisions for translating the specification into user manuals, developing software interfaces that can be easily reconfigured for different categories of users, and writing the interfaces in the style and language of the user environment.

The specification should further cover the packaging, publications, and installability considerations for the product. Then, of course, there should be detailed sections for each of the software quality attributes

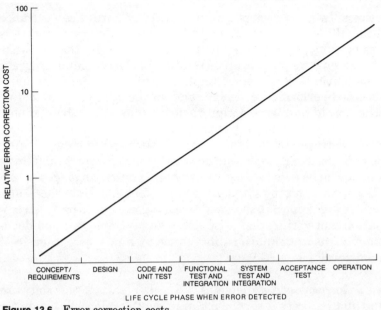

Figure 13.6   Error correction costs.

listed in Figure 13.2. Lastly, the specification should include constraints around cost, schedule, and any special trade-off considerations.

13.6.2   Performance of software reviews and audits

From a software quality assurance viewpoint, this set of activities is extremely important. The earlier potential errors are detected, the easier and less costly it is to correct them. Figure 13.6 shows this classic relationship very well.

Specification requirements review.   A review of the specification requirements is essential for all the reasons stated above. The review must be done carefully and thoroughly to ensure completeness and adequacy. The specification requirements must be measurable and verifiable, cover each of the software quality attributes, and provide the user perspective. Any shortcomings in any of these areas should be corrected immediately, since allowing them to persist is asking for trouble. The specification requirements set the direction for the development effort. The wrong direction will result in later deleterious cost and schedule impacts.

Design reviews. A trusty weapon in the arsenal of hardware assurance has traditionally been an effective design review program, and it is equally important to software quality assurance. In the software domain, these reviews provide an evaluation of the computer program design to establish if the proposed implementation is capable of meeting its specified performance, design, and verification requirements.

As in the case of hardware design reviews, they should be planned, documented, and action items assigned, followed up, and closed out. These design reviews should examine items such as the programming technique to be used (e.g., top-down, bottom-up, modular programming); coding practices to be followed (e.g., use of higher order languages, structured coding, programming standards); major functions, interfaces, and data structure breakdowns to be considered; algorithms to be constructed; validation and verification approaches to be employed; and documentation (e.g., user manuals, maintenance manuals, acceptance test plans) to be developed.

Code reviews/inspections/walkthroughs. Another part of the software review and audit process is to conduct an in-depth review of a computer program, or a portion of a program, by inspection. This type of review is done by a team of peers who walk through the program and examine it intimately, looking at it from diverse viewpoints in an attempt to find errors which may have escaped the programmer. Although this process is slow and tedious, it has a high payback as errors are caught early.

It should be noted that code "inspectors" need not necessarily be from the software quality assurance team. Although the peer team should include technologists who have a direct interest in the program (e.g., part of the same programming team), it should exclude managers. In addition, if this process is to be successful, the team should adopt the attitude of being tactful, cooperative, and nondefensive.

Code inspections should be aimed at finding problems such as those due to coding logic, code comments, design error, interfaces, performance, language use, standards, storage use, flowcharting, and documentation. These inspections should be performed after coding of the program is complete, well annotated, syntactically correct, and free of rudimentary errors as a result of the first successful compilation of the program (but not necessarily debugged).

Other review/audit approaches. Surely there are other useful functional, physical, and in-process reviews and audits of the software development which can, and should, be conducted to assure that it is proceeding in the right direction toward meeting the requirements. Basically, these reviews and audits are associated with the application of various software tools and techniques which serve to ferret out software problems, which once identified, can be corrected. Some examples of the kinds of tools and techniques which should be monitored from a software quality assurance viewpoint for both implementation and results are briefly described here.

The design review process certainly provides some insight into the chances of the design being correct when completed. Another tool is the use of an automatic design checker. The checker is essentially a computer program which accepts the representation of the design as an input, analyzes that design, and then provides a listing of design flaws. However, the key to employing this tool is that design representation be sufficiently formal to allow its use.

Another well-recognized technique in striving for program correctness is manual checking. Certainly less sophisticated than automated checking, manual checking is just what it says—checking out the program totally by manual means. The program can be reviewed for faults; arithmetic calculations can be done to verify the correctness of outputs; and program flow and data flow can be understood and verified. To raise the level of effectiveness of this activity, efforts can be concentrated on special problem areas such as code inefficiencies and suspected errors. Thus, manual checking still has its place in the assurance arsenal.

A couple of other audit areas might include the structural analysis and correctness proof evaluation. Through the use of the former technique, a structural, automated analysis of a computer program is performed to evaluate its data and logic structure, and to look for deficiencies such as undeclared data variables or missing logic. Via the latter method, mathematical theorem-proving concepts are applied to the program or its design to verify consistency with the specifications. Again, both audit areas provide an opportunity to find and fix software problems, and audits should be performed to make sure that they are doing just that.

13.7 Documentation

Software documentation, including user and maintenance manuals, should also be reviewed for adequacy as part of the software quality assurance program. All relevant documentation should be thoroughly reviewed to ensure that it satisfies accurately and completely the performance and design requirements of any higher level documentation. In addition, the documentation should be further verified for accuracy and completeness by checking for all components, correct cross-referencing, and editorial accuracy. These reviews should be performed on both an informal working-level and a formal basis. The formal reviews should follow the working-level reviews so that changes can be incorporated and a corrected version of the documentation provided for the formal reviews.

13.8 Analytics

In the analytical area, a variety of tools and techniques has been developed for use in the software quality assurance arena, primarily for application in the software reliability area. In the context of software, "reliability" was previously identified as an important quality attribute and defined as "the extent to which (the) software can be expected to perform its intended function with required accuracy." These tools and techniques provide a way of measuring software reliability.

The sophistication of analytical tools/techniques available to the software quality assurance analyst ranges from the very simple to the highly complex. In this section, the more commonly used, and proven, approaches are discussed rather than those which are more "experimental" in nature. However, it is not intended that the discussion be an in-depth treatment. Rather the intent is to demonstrate to the reader that there is in fact a wealth of tools and techniques which he or she can select from to address the issue of software reliability measurement.

In this discussion, two important words, "failure" and "fault," surface repeatedly. To eliminate confusion, we apply the following definitions:

- *Failure.*   The inability of a functional unit, component, or system to perform its required function within specified limits.

- *Fault.*   An accidental condition manifested by a software error (due to human action) which, if encountered, may result in a failure.

13.8.1 Fault density estimation

Since residual faults can result in software failures, they are obviously undesirable. Thus, it is important to have some insight into the number of faults remaining in the software. One way to do this is by estimating the fault density FD using the simple relationship:

$$\text{FD} = \frac{N_F}{\text{PS}}$$

where $N_F = \sum_j nj$ or the total number of faults found, (where nj is the number of faults found per the jth failure), and PS is the program size, usually expressed in thousands of source lines of code.

Each fault can be classified as to severity and class. The estimated fault density can be compared with the expected fault density. In addition, observed faults can be compared with goals for the severity classes. From these comparisons, determinations can be made as to whether sufficient testing has been accomplished for that program type and size.

13.8.2 Cause and effect graphing

Poor specifications are notorious for causing software quality problems. Fortunately, a graphing tool is available to help to determine if software specification requirements are ambiguous and incomplete. This technique, called *cause-and-effect graphing*, examines program inputs and expected outputs, and serves to surface the ambiguities. All the existing requirements are identified, broken down into separate segments, and analyzed to pinpoint the various causes and effects in the specifications.

The cause-and-effect graph is constructed by proceeding through the following steps:

1. Assign and uniquely number a node associated with each specification cause and effect.

2. Transform the interconnected nodes into a boolean graph through analysis of the specification semantic content.

3. Set the possible true-false states of the various causes and establish the conditions related to each effect.

4. Identify the cause-effect combinations on the graph for those constraints which cannot occur because of syntactical or environmental reasons.

5. Flag as ambiguities (a) those causes which have no corresponding effect, (b) those effects which do not originate with some cause, and (c) any combination of causes and effects which is either unachievable or inconsistent with the specification requirements.

From the graph, an index of relative ambiguity I_A can then be determined from the simple relationship:

$$I_A = 1 - (A_R/A_T)$$

where A_R is the number of remaining ambiguities which still need to be eliminated and A_T is the total number of ambiguities identified through the cause-and-effect graph. Obviously, as the ambiguity index approaches 1.0, the number of residual ambiguities goes to 0, indicating that all existing ambiguities have been eliminated.

13.8.3 Minimal test case evaluation

In the interests of testing efficiency, it is useful to determine the minimal number of test cases that need to be run. This determination can be made by evaluating the complexity of the software module and establishing the number of distinct paths through it. Again, a graphing technique helps facilitate the analysis. Each path through the module is identified and graphed, and test cases are constructed along each path traversing all the edges of the graph. The test cases, when executed, presumably cause all the statements and related decision outcomes to be exercised a minimum of one time each. The result is interpreted as the minimal number of cases which provide test coverage of all paths through the software module.

13.8.4 Run reliability probability estimation

An important probabilistic measure which the software quality assurance analyst can calculate is one which estimates the probability that a randomly selected run will yield correct results. The technique used is called the Nelson method, so named for E. C. Nelson, who developed it while at TRW (One Space Park, Redondo Beach, CA) in the early 1970s.

A random sample of inputs is selected from a defined set of all possible discrete input patterns. Weighted probability values are assigned to each possible input pattern. The randomly selected number of trial runs N_{TR} is established consistent with the desired estimate accuracy, with each run having a probability p_i. The probability of a successful run P_{SR} can then be estimated as

$$P_{SR} = \frac{P_{CR}}{P_{TR}}$$

where P_{CR} is the sum of the probabilities of all the correct runs and P_{TR} is the sum of the probabilities for all runs or, in the case of a uniform distribution,

$$P_{SR} = \frac{N_{CR}}{N_{TR}}$$

where N_{CR} is the number of correct runs.

The estimate, P_{SR} can be further used to determine the probability P_{ASR} that a given number of runs m will all yield the correct results

$$P_{\text{ASR}} = (P_{\text{SR}})^m$$

One caution that should be given here, however, is that the system must be in the same state at the start of each run.

13.8.5 Mean-time-to-remove-remaining-faults estimation

There are any number of existing models (e.g., Jelinska-Moranda model, Schick-Wolverton model) which are comprehensively described in the literature and which are based on the time-between-software failures. Consequently, they will not be covered here. Nevertheless, the reader should be aware that these models are useful for determining the average (mean) time (e.g., central processing unit time) to detect the next j faults. The importance of the resulting measure is that an estimate can be established for the additional time necessary to reach a desired software reliability level.

13.8.6 Remaining faults estimation

Closely associated with estimating the mean-time-to-remove-remaining-faults is the estimation of the number of faults remaining in a program. One method of deriving the latter estimate is through the technique of error seeding (see "Development Testing" later in this chapter). This technique involves randomly inserting into the program some number of faults n_s which are typical of the expected innate faults. The distribution, quantity, and types of seeded faults are determined from a fault analysis.

The program is then tested for some predetermined period and monitored; then its identified faults are classified as being seeded or indigenous. It should be noted that the fault search is conducted by individuals who have no knowledge of the seeded faults. From the resulting data, the number of remaining innate faults can be determined by type as

$$\hat{N}_R = (n_u n_s)/n_f$$

where $\hat{N}_R$ = the maximum likelihood estimate of the number of remaining indigenous faults of a specific type truncated to the integer value
n_u = the number of found unseeded faults
n_s = the number of seeded faults
n_f = the number of found seeded faults

As in any sampling scheme, the number of seeded faults significantly impacts the degree of statistical confidence in the estimate (see Assess ment in Chapter 9, Analytics). Therefore serious consideration should be given to both the sample size and the representativeness of the inserted (seeded) faults.

13.8.7 Reliability growth estimation

As in the case of hardware-oriented reliability growth testing (see Reliability Growth Testing in Chapter 11, Testing), a strong parallel exists in the area of software reliability growth. Software fault density or failure rate goals can be established and testing performed to find faults and correct them. As in the case of mean-time-to-remove-remaining-faults, there are several models available and described in the literature (e.g., Goel-Okumota model) which can be used for estimating when the established goals will be reached.

Additional parallelism with hardware-oriented reliability growth testing exists in the way software failures and associated faults are processed. All software failures are carefully recorded as to time of occurrence, conditions of failure, failure symptoms, etc. In addition, comprehensive failure analysis is performed to determine what corrective action should be taken. More importantly, however, a careful analysis is undertaken to ascertain whether the failure occurred because of a previous corrective action attempt—a not uncommon situation in the case of software fixes. Thus, the rate at which faults are being reinserted during the corrective action process should be well understood to gain a proper interpretation of the reliability growth function.

13.8.8 Mean-time-to-failure estimation

The choice of models for estimating the mean-time-between-failures (software) is extensive and is dependent on the time-to-failure distribution and associated assumptions. Prior knowledge of this failure distribution will aid significantly in selection of an appropriate model. The most basic of these models simply takes the time t_i between the ith and the $(i-1)$st failure and assumes a constant failure rate and an exponential distribution function (see discussion on distribution functions in Statistics in Chapter 9, Analytics). A finer breakdown can be made by classifying the software failure as to severity and deriving mean-time-to-failure (MTTF_s) estimates for each severity class. Thus, this simple model translates mathematically to the following:

$$\mathrm{MTTF}_s = \sum_{i=1}^{j} \frac{t_{ij}}{f_j}$$

where t_{ij} is is the time-between-failures for the jth severity failure class and f_j is the number of failures of the jth severity class.

Again, detailed record keeping is required to develop the database of time-between-failures and number of failures, supplemented by comprehensive failure analysis to arrive at severity class determination. Also, as in the case of reliability growth estimation, the impact of reinserted faults must be well understood. These concerns notwithstanding, the resulting MTTF can prove useful for comparison against specified values.

13.8.9 Other analytical techniques

In addition, to the techniques covered here, there are many other commonly used, proven methods for evaluating the number of conflicting software requirements, verifying the traceability of requirements, assessing the simplicity of a software program, estimating the failure occurrence rate, analyzing the relative difficulty of software maintenance, examining the degree of test coverage, etc. Thus, there are tools which the software quality assurance analyst can use for measuring both the goodness of the software development process and the software end product.

13.9 Testing

Testing is an essential step in the process of assuring the quality of software. Toward this end, there are several kinds of tests which are employed: development tests, field tests, regression tests, software quality tests, and acceptance tests. These tests are described in the paragraphs that follow. Additionally, insight is provided into the role of the software quality assurance function in regards to testing.

13.9.1 Development testing

This category of tests covers the application of various development test tools, as well as the implementation of specific, more formally staged, development tests. Both types of activities are discussed in this section.

Development test tools. An extensive set of test tools is available to aid the developer in developing quality software. The application of these test tools should be monitored, together with the results, as part of the software quality assurance program, to ensure that identified problems are corrected on a timely basis. An overview of some of the more commonly applied development test tools is as follows:

- *Source language debug.* Preplanned source language debug statements are inserted in a program to aid in its debugging in the language in which it is coded, rather than in the machine language of the computer.

- *Test coverage analyzer.* A "special" computer program is applied to the program of interest, and a count kept as to which logic paths are tested (and obviously which are not tested) and how many times they are executed.

- *Assertion checker.* A modified version of the source code is executed to check out statements (assertions) which are presumed true and to detect violations.

- *Environmental simulator.* Inputs are provided to the software or hardware/software system and processed with the appropriate interfaces exercised under a scenario which attempts to represent the "real world."

- *Test data generator.* Test cases are generated via computer program in an attempt to generate an optimum set.

- *Test driver.* A "special" computer program tests another computer program or its components in a bottom-up fashion, so that as each component is validated, it is added to the software structure of which it is a component.

- *Error seeding.* Software errors are deliberately introduced into the program so that a determination can be made of the number of errors detected during debugging, and the results are used to develop estimates and establish confidence levels of existing errors and the time required to discover them.

- *Interactive debug.* Computer program errors are sought and corrected while communicating with the computer executing the program via a time-sharing system or single-user hands-on approach so that the progress of the program is monitored, intermediate values are inspected, and data corrections are inserted.

- *Foreign debug.* The program is debugged by someone other than the original programmer via the use of a peer code review discussed earlier, or via a test in which the disinterested parties develop and execute various test cases.

- *Mathematical checker.* Mathematically based programs are checked for errors using a process which evaluates the accuracy or the significance (i.e., number of significance digits or bits of accuracy) of a computerized mathematical solution.

In a more formal sense, the software quality assurance group should become intimately involved in monitoring specific development tests, namely module tests, subprogram tests, program performance tests, and system integration tests. This monitoring should include not only the conduct of these tests, but also the planning and evaluation.

Test planning. At the outset, the software quality assurance group should ensure that the right tests are being planned to verify that the software meets all the operational and technical requirements, as well as the established acceptance criteria. The test plans should describe in detail what levels of testing will be conducted to verify that the performance requirements have been met; what the specific criteria are for accepting the software; what the detailed procedures are for performing each level of test; organizationally who will be participating in these tests and what their roles and responsibilities are; where each test will be performed; what the test schedules are; and what results will be reported and how they will be processed.

Module tests. Presumably, each software module will have previously proceeded through up-front assurance activities (see "Design"), such as code walkthroughs, prior to any testing. Module tests will then be conducted to ensure that each coded module can be compiled or assembled without errors. The coded modules should be fully exercised to verify that they fully satisfy the requirements established for their design and performance. During these tests, required inputs and outputs should be consistent with the established requirements.

Subprogram tests. After having successfully completed the module level tests, the modules will be linked and integrated individually into segmented subprograms and subjected to testing at that level. Again, these tests must be sufficiently adequate to determine that the established operational, performance, and technical requirements have been satisfied. The software quality assurance group should ensure that the modules have been linked and integrated without error. As in the case of the module level tests, the various inputs and outputs to the subprograms must be fully exercised to verify that the appropriate design and performance requirements have been met. As part of this exercise, erroneous inputs should be provided to prove the ability of the subprograms to handle them properly and maintain their integrity. In addition, the subprograms should be checked for user friendliness.

Program performance tests. Successful subprogram tests will then be followed by program performance tests which provide a way to individually integrate the subprograms into the program and to verify compliance with the overall requirements. Some of the performance characteristics which should be verified include the ability of the program to initiate and restart the system, to load the program, and to make data entries through peripherals. All interfaces, both human-machine and equipment-equipment, should also be checked, as should the ability of the system to handle erroneous inputs and still function properly.

System integration tests.   If the software being evaluated is a piece of a larger system which will be merged with other software, or networked, then the successful integration of the software segments should also be verified.

The above development-oriented tests, together with the up-front design activities (e.g., peer code reviews), will allow us to go a long way toward wringing out potential problems from the software. Active participation by the software quality assurance group will ensure that "things don't fall through the cracks" at this important stage in the software development.

13.9.2   Field testing

As part of the overall test program, software testing is also typically conducted under actual user conditions in the field. The in-house testing takes the specification requirements and attempts to verify that they have been satisfied. However, as noted at the outset, in the final analysis, it is the user of the software product, including its associated documentation, who will ultimately have to be satisfied that the software does what it's supposed to do, error-free. Consequently, the concept of field testing (or so-called beta testing) takes on significant importance in the process of software quality assurance.

Field testing is usually accomplished at a number of "friendly" customer sites to gain the benefit of the direct user perspective. It provides a twofold benefit: (1) the customer has an early chance to try out the software and documentation, and provide some inputs for enhancements and (2) the developer has the opportunity to take this feedback and make changes and modifications before the software and documentation are formally released to the "world."

Exercising the software product under user conditions allows for exposure to inputs and events which may not have been fully imposed during in-house testing. Thus, the product can be judged more realistically. However, the major caution, of course, is that the selected field sites be representative of the overall potential customer base. Otherwise, the results may be significantly biased.

Finally, an essential part of the field test activity is the evaluation and corrective action process which takes place on the reported results. The results will only be useful if they are acted upon in a timely and effective manner. As such, field test results should be subjected to the improvement activities discussed later in this chapter under "Improvements."

13.9.3 Regression testing

Too often in the case of software changes, a supposed fix or modification causes some other kind of problem which the designer or programmer never dreamed would happen. In an attempt to prevent such a happenstance, the technique of regression testing is employed.

Associated with software maintenance, regression testing is directed toward detecting errors in changes or errors caused by changes. Regression testing is aimed at ensuring that changes are correctly made, that the original problem is indeed corrected, and that no other problems are introduced in other parts of the software as a result of the change.

Regression tests should take advantage of test approaches used during acceptance testing. The changed version of the software should be subjected to a well-defined set of tests which utilize acceptance kinds of tests, as well as tests which are peculiar to the specific change(s) made.

13.9.4 Software quality testing

The aim of software quality testing is to assure, to the maximum extent possible, that there are no critical "bugs" remaining in the software and that it is relatively error free. To this end, the software quality assurance group performs testing to make this determination. In such testing, the software is stressed beyond the limits of its design capacities, and then some, to verify that degradation will not prove fatal at the point of saturation.

The duration of the test should be established consistent with the expected usage of the product so that all inputs, functions, and interfaces are fully exercised and stressed in the way the product will be used. Once the software to be tested is loaded, initialized, and started, the testing should run continuously for a preestablished test duration. Both normal and abnormal inputs should be applied and carried out in some random pattern to check variations in operational as well as in normal modes. In addition, for shorter periods of the test, the capabilities of the software should be stressed in terms of data handling capacity and response times to the point of saturation by exceeding data volume and data rate limits, e.g., exceeding processor handling requirements and saturating data transfer capabilities.

The software quality test criteria should clearly define the allowable number of errors. These criteria should be specified as maximum limits and related to error severity as follows:

- *Critical.*   Prevents the performance of an operational function that is consistent with established requirements

- *Major.*   Affects adversely the performance of an operational function that is consistent with established requirements so that there is performance degradation with no work around possible

- *Minor.*   Affects adversely the performance of an operational function that is consistent with established requirements so that there is performance degradation but for which there is a work around

- *Nuisance.*   Causes user inconvenience with no impact on operational function

Generally speaking, the occurrence of critical or major errors should be considered as an unacceptable test outcome that requires both correction and test rerun with no patching allowed.

13.9.5   Acceptance testing

After successfully completing software quality testing (i.e., satisfying preestablished acceptance criteria for unresolved software/documentation errors and patching), the software can be submitted for acceptance testing. If there is a distinct outside customer, acceptance testing usually takes place at the customer site during software delivery and installation. If not, acceptance testing may be conducted by a group independent of the development organization and monitored by the software quality assurance function, or by the latter group directly. In either case, acceptance testing of the software should be performed in the ultimate user environment for which the software is designed, whether on-site or not. As part of the acceptance tests, all of the software functions should be exercised for some predetermined time period which allows for latent errors, if present, to appear.

13.10   Improvements

Improvements to software can be precipitated by two primary processes: (1) correction of software quality problems or (2) enhancement of software performance (e.g., greater speed) or features (e.g., more functionality). Software improvements can be patches or work arounds to a problem, or upgrades in which the software is revised or modified, and a new version or release issued. The patch is usually a quick fix to the problem to tide the user over until the more permanent upgrade can be issued. Depending on the severity of the problem, a patch may be appropriate or a new release may be necessary. In the case of enhancements, the competitive pressures of the marketplace will usually dictate the timing of the release.

13.10.1 Software problem reporting, analysis, and corrective action

Problem correction should not be limited to the use phase. Since opportunities for improvement occur over the entire spectrum of the software life cycle, it is important that such a process begin early in the development phase and continue through the use phase. And the heart of any improvement effort, whether hardware or software directed, is an effective problem-reporting, analysis, and corrective action process. Such a process can ensure that problems are corrected and do not slip by; that progress in error correction is tracked; that feedback is provided on the status of error or trouble correction; and that proper priorities are established to fix reported errors and problems.

One dilemma that has to be resolved in any formal software tracking process, however, is the timing for initiating the process. Surely, there are significant quantities of errors that are discovered during the design and debugging stage (e.g., as related to specification deficiencies), and also during the code and checkout stage, that should be tracked formally and corrected. However, because of the large number of errors, initiation of the tracking process is sometimes deferred until later in development testing (e.g., when system integration testing begins). Nevertheless, since the benefits of early initiation far outweigh the cost (refer to Figure 13.6), the process should be started in the design phase, even if implemented on a less formal basis during that stage.

Software problem reporting. Software problems should be reported on hard or soft copy prenumbered forms, which minimize the chance of "misplacing" reports and not knowing about it. A sample software problem report (or SPR) form is shown as Figure 13.7. The form should allow for entry of essential information such as problem symptoms, analysis and closure status, problem originator, correction priority, problem classification (e.g., documentation error) problem and fix verification, etc. After the initial entries are filled in, the problem should be referred to the responsible software organization for analysis.

Software problem analysis. This is where the fun begins! The key to correcting a problem is determining its cause. And the analysis for cause determination becomes a function of the available background information (e.g., symptoms and conditions surrounding the problem) and the competency of the analyst. Even a competent analyst will have difficulty in analyzing a problem if meager information is on hand or obtainable. Conversely, complete information in the hands of an incompetent analyst will not guarantee a fix either. The analysis is made more difficult because software problems are usually varied and complicated, and often are not duplicative. As a result, software problem analysis typically involves much detective work to track down the cause, and for the more difficult problems, the use of a trial-and-error approach to establish the probable cause. In any event, after the analysis is done, the results are entered on the SPR.

Software problem corrective action.   Once the analysis is completed and the probable cause determined, corrective action can be initiated. It is at this stage in the process that the software quality assurance function must be especially watchful. Too often a supposed fix causes another problem elsewhere in the software because faults are reinserted. To prevent this from happening, we should, after the change is approved and the correction coded and inserted in the program, first execute the software against all possible failure conditions and then against a regression test (see previous section on regression testing). Should any of these tests fail, it's best to cycle through again—correction, redesign, recode, and retest. If successful, the software change can be released. Also, another caution, if there is an undisciplined process in place, a corrective action may be proposed, supposedly but not actually implemented, and the problem closed out only to reappear again. The stop-gap measure in this instance is close monitoring and follow-up.

Software problem database.   Software problem data can provide a wealth of information relative to the status of various reported problems. There will be information on old problems and the resolution status of existing problems classified as open, pending (corrective action), or closed. For unclosed problems, there will be insights into probable closure dates. For new problems, probable disposition can be established. Patterns and trends can also be determined for problem causes, open SPRs, etc. Last, the database can provide a history file against which to check error removal rates and previously encountered problems.

13.10. 2 Configuration management and change control

Another important software quality assurance audit area is the configuration management and change control system employed. The objective of this system should be to ensure the positive identification,

SOFTWARE PROBLEM REPORT

Problem Report No. _____101_____

Project Name _____ Computer _____ Program _____

| Problem Report | Reporter _____ Date _____ |

Where Detected
☐ Requirements Review
☐ Design Review
☐ Code Review
☐ Module Test

☐ Subprogram Test
☐ Program Performance Test
☐ System Integration Test
☐ Field Test

☐ Regression Test
☐ Quality Test
☐ Acceptance Test
☐ Operation

Description of Symptoms / Problem _____

_____

Configuration Level _____
Corrective Action Priority / Date _____
Approval Signature _____ Orgn. _____ Date_____

| Problem Analysis | Analyst _____ Date _____ |

Analysis Results _____

_____

_____

_____

Approval Signature _____ Orgn. _____ Date_____

| Problem Correction | Corrector _____ Date _____ |

Description of Corrective Action _____

_____

Changes / Configuration / Level _____

Problem Category
☐ Specification
☐ Coding

☐ Design error
☐ Testing
☐ Configuration management

☐ Documentation
☐ Other

Approval Signature _____ Orgn. _____ Date_____

Figure 13.7 Sample software problem report (SPR) form.

control, and status accounting of the software configuration. The system should address both the software program and associated documentation.

Configuration identification. Internal software baselines should be established to represent the then current, identified software configuration. These baselines should be set up at appropriate internal departure points which are necessary to track future changes in performance, design, and associated technical requirements. In this way, the baselines provide for an orderly transition between the various software development stages. The same sort of identification concerns should also be applied to the documentation as well. Software documentation should be titled, labeled, sequence numbered, dated, and cataloged to provide proper identification. The documentation should be related to the specific software component to which it applies; the software baseline of which the documentation is a part or supports should be defined; the edition and change status of the documentation should be denoted; and, of course, the purpose of the documentation should be described.

Configuration control.   Appropriate procedures should be developed to control formally all software program materials and documentation and to include the operation of a development support library. Records of program modifications should be carefully maintained in the library to provide audit trails for these updates. Information on date and description of last change, as well as the initiator of that change, should be included in these records. Modules with many updates should be flagged and evaluated for possible redesign.

All proposed software changes should be evaluated for interface, schedule, and cost impact, and if approved and implemented, placed under configuration control. Similarly, changes to documentation should be put under control procedures as well to ensure that only approved and descriptive changes are implemented and that they are done so on a timely basis. Changes should be formally classified as to whether they affect functional or allocated baseline specifications. An additional informal change classification should also be established for development changes which occur between unit release and product baseline without any impact on baseline specifications.

The most efficient way, perhaps, to maintain effective control is through the establishment and operation of a software configuration control board (SCCB) which formally controls any change to these baselines. The SCCB should review and evaluate all proposed changes (e.g., for technical design impact) and for approved changes, ensure timely implementation on all baseline documentation. As part of the overall control process, the software quality assurance function should perform audits at key points in the development and verify that appropriate procedures are being effectively implemented so that control objectives are attained and maintained.

14

Careers

*No man can better display the power of his
skill than in so training men that they come
at last to live under the dominion of their
own reason.*

SPINOZA

14.1 Directions

To many individuals, the choice of a career is one of the most important decisions they will face. It will affect their marriage, retirement,
standard of living, happiness, and even their personality. Whatever
career they choose, it should fulfill their needs whether they be intan-
gible (e.g., job satisfaction) or tangible (e.g., money).

Just as the assurance sciences have increased in importance, so too
have related career opportunities. A qualified individual can seek product assurance employment in every general business area, i.e., government, industry, consulting, and education.

Aside from the increase in importance of the assurance sciences, the
technical advances that have occurred in the disciplines as they have
matured have created increasingly diverse employment opportunities.
Examples of how technology advances have increased career potential
are almost as numerous as the number of technological advances during
the same time period. Some significant ones are explained below.

*The exponential expansion of the use of computers for data retrieval
and analysis* has created career paths for personnel interested and
trained in computer applications into all areas of product assurance.
Computer operators and specialists are found in reliability and maintainability organizations—computer circuit analysis, system analysis,

and reliability/maintainability/availability predictions; in quality organizations—software quality assurance management, visual inspection
data analysis, test data analysis, and computer-driven test equipment;
and in components engineering organizations—development of data-
banks of parts application and history data.

The tremendous advances and growth in electronic parts has
presented a whole new set of assurance careers for people with backgrounds in physics (failure analysis), semiconductor design (components
engineering), process development (process control), etc.

*The almost unbelievable technological advances in large systems
design and development* has spawned whole new specializations in the
design assurance sciences utilizing many mathematically trained and
specialized personnel (analysis techniques such as queuing theory,
Monte Carlo simulation, bayesian statistics, etc.).

This chapter is devoted to (a) identifying how one acquires the qualifications needed for a career in product assurance, (b) describing the
diverse career employment possibilities, (c) examining briefly the
rewards offered, and (d) "forecasting" what lies ahead for one seeking
a product assurance career.

14.2 Qualifications

As we have noted throughout this book, the product assurance disciplines involve activities such as planning, analysis, and testing. To
execute these activities, there is a need for technical and nontechnical
personnel possessing a variety of backgrounds. For example, there is a
need for people with engineering and mathematical/statistical analysis capabilities, as well as for those with planning abilities and management skills. There is a definite demand for a wide variety of product
assurance practitioners as attested to by the "Help Wanted" sections
of most major newspapers. So how does one acquire the proper credentials to enter the product assurance field? In order to answer this question, we have to look at what the qualifications of a product assurance
practitioner are, as well as how they are attained.

The qualifications for a career in product assurance are as diverse
as the disciplines involved in assuring product integrity. Academically, these qualifications range anywhere from nondegreed personnel to
individuals with doctorates. In terms of experience, they cover the
gamut from zero to many years.

There are technicians: failure analysis technicians, test technicians,
process control technicians, quality control technicians, and field service
technicians—to name a few. The qualifications necessary for these jobs
are the normal electronic, mechanical, chemical, or computer techni-

cian training. Specialized skill training in the assurance disciplines for technicians is generally unavailable except through work experience. This limits the number of experienced technicians in the assurance field and generally enhances their value.

Clerical personnel with statistical training are needed in many assurance disciplines. Personnel with computer programming experience are also frequently utilized. General administrative personnel often find a rewarding career in the assurance field.

Likewise, hourly manufacturing operators and production workers who develop considerable skill at their particular tasks find—sometimes through additional training—quality inspector jobs. These inspection jobs allow for continued growth, with additional experience and training, into either supervisory roles or more highly skilled quality specialty careers such as source inspectors, process control specialists, quality auditors, and sometimes quality engineers. This same growth paths frequently exists in other areas of manufacturing such as production control. It is of significance that this career movement is generally a growth move in that higher skills are needed and, therefore, higher compensation so provided. Note particularly that the nonprofessional assurance careers are invariably developed almost entirely through work-related experience and training in a more general occupation.

There are three basic routes for gaining these qualifications and hence a career in product assurance. First, there is the academic route. An individual acquires the appropriate academic training and obtains a product assurance position which exploits this training. A second route is through some related work experience that provides a useful background which can be applied in product assurance. The last route is a combination of the two—pertinent academic training and relevant work experience. These career routes are examined in more detail below.

14.2.1 Academic route

The academic route covers more than just college and university study. It also includes civilian and military technical schools. Don't necessarily look for colleges or schools offering a curriculum leading to a degree or certificate in product assurance. These are the exception rather than the rule. Instead, look for institutions which offer *pertinent* academic training. What then is considered to be pertinent?

Table 14.1 provides a matrix cross-referring the more pertinent college majors or certificate-type specialties to the product assurance disciplines. As can be seen from this matrix, the basic engineering majors and specialties are prime requisites to most of the product assur-

TABLE 14.1   Pertinent Educational Specialties vs. Product Assurance Disciplines

Educational specialty	Product Assurance Disciplines					
	Reliability	Maintainability	Human factors	Manufacturing quality assurance	Safety	Software quality assurance
Mechanical engineering	X	X	X	X	X	
Systems engineering	X	X				
Computer science/engineering	X					X
Physics	X					
Mathematics/statistics	X	X		X	X	X
Psychology			X			X
Electronic/electrical engineering	X	X	X	X	X	
Manufacturing engineering				X		
Engineering management	X	X				
Industrial engineering	X	X		X		

ance disciplines. Since product assurance is engineering-oriented, this should come as no surprise. Other types of academic training are also useful, particularly when supplemented by appropriate work experience. The sources of such training cover many of the major colleges and universities (e.g., University of California at Los Angeles), civilian schools, and military schools (e.g., Navy Electronics School).

There are also a few institutions which offer direct academic training in the product assurance disciplines. Northeastern University, in Boston, offers state-of-the-art courses in reliability engineering, quality control, and product assurance; the University of Arizona has a master's degree program in reliability engineering; Texas A&M University, in conjunction with the U.S. Army Material Command, offers a master's degree program in maintainability engineering. In addition to these formal approaches to learning are informal training courses. They may be offered or sponsored by colleges or universities, nonprofit organizations, professional societies, and the military. For example, George Washington University offers short courses in quality management and reliability and systems safety engineering; the University of Arizona annually sponsors a two-week institute in reliability engineering. ARINC Research Corporation regularly conducts training sessions in reliability and systems effectiveness; the Air Force sponsors courses in statistical quality control; the Institute of Electrical and Electronics Engineers (IEEE), through its Professional Group on Reliability, offers courses in reliability at the local chapter level; and the American Society of Quality Control (ASQC) runs courses on various aspects of quality control.

In addition to these informal training courses, there is a whole field of home "correspondence" type courses—many of which deal directly or indirectly with the assurance sciences. "Programmed learning" courses in reliability engineering, system effectiveness, failure analysis, and statistical quality control are just a few of the wide variety available. These courses are well developed and more easily utilized than their more formal academic counterparts.

Thus, the academic route has many branches which include both formal and informal training in pertinent or directly applicable course matter. They can all lead to a career in product assurance.

14.2.2 Work experience route

"There is no substitute for experience" is an often heard statement. This is also true with respect to the product assurance disciplines. Different kinds of work experience provide a necessary "passport" into the product assurance field.

From the outside looking in, the type of experience considered to be the most relevant depends on where our immigrant wants to roost. As you may recall, the product assurance responsibility can be subdivided into two major areas of concern—one involving design assurance and the other involving manufacturing assurance. Related to the former are the reliability, maintainability, systems safety, human factors engineering, and software quality assurance disciplines, as well as associated disciplines such as systems effectiveness and integrated logistics support. Related to the latter is the traditional manufacturing quality assurance discipline. The most pertinent credentials for entering the design assurance disciplines are work experience in design and development oriented work. Similarly, the "best" experience for the manufacturing quality assurance side of product assurance is manufacturing, industrial engineering, and/or test experience.

However, we shouldn't limit our relevant work experience to design, development, manufacturing, and test. Although these areas are certainly the most pertinent, there are other types of experience that are extremely helpful for entry into the product assurance field. Table 14.2 provides a matrix showing the various product assurance disciplines and the more common types of work experience considered to be useful.

So far, we have only looked from the *outside* into the product assurance area. We can also look *within* product assurance for relevant work experience which would permit transfer from one product assurance discipline to another. Figure 14.1 shows some very viable possibilities. Since reliability is a "birth-to-death" kind of discipline in terms of responsibility, there would be coverage and concern for quality problems; therefore, someone with reliability experience would adapt easily

TABLE 14.2   Related Work Experience vs. Product Assurance Disciplines

	Product Assurance Disciplines					
Educational experience	Reli-ability	Maintain-ability	Human factors	Manufactur-ing quality assurance	Safety	Software quality assur-ance
Manufacturing				X		
Design engineering	X	X	X		X	
Project management	X	X	X	X	X	X
Test				X		X
Software engineering						X
Systems engineering	X	X				X
Operations research	X	X				X
Industrial engineering	X	X	X	X		

into the quality assurance function. Maintainability engineering is involved with the ease of repair, and human factors engineering considers the physical characteristics (e.g., arm reach) of repair personnel. Software quality assurance is concerned with both the reliability and maintainability of the software. Safety is concerned with potential damage to the equipment (reliability) and to humans (e.g., during maintenance). There is much potential interplay between the product assurance disciplines.

14.2.3   Combined academic and work experience route

This last route is perhaps the most obvious. We have described the kinds of academic training that are useful for a career in product assurance.

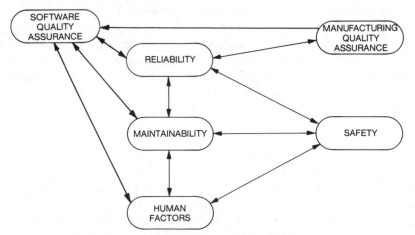

Figure 14.1   Transfer possibilities within product assurance.

Similarly, we have discussed the types of work experience that can be pertinent to a product assurance career. If we bring these two approaches together, we have many mixes of academic training and work experience that can be used to create a product assurance practitioner. However, some combinations are more desirable than others, as shown in Table 14.3.

A review of Table 14.3 reveals a most interesting observation—academic training in one of the engineering specialties, supplemented by work experience in mathematics/statistics, provides solid product assurance credentials. And when you think of it, why shouldn't they? Product assurance, as we said from the outset, is a science. It requires a strong knowledge of engineering *and* the ability to measure results.

14.2.4 The product assurance generalist

To this point, we have discussed product assurance qualifications only in terms of their individual disciplines. As you gain the proper credentials, you become a reliability engineer, quality engineer, or whatever.However, a particular breed of product assurance professional is the product assurance generalist.* This individual must know all aspects of product assurance and how they tie together, yet does not get bogged down in details. The product assurance generalist is a rare breed, whose qualifications include the following:

1. Working knowledge of the techniques/principles involved in the various product assurance disciplines
2. Ability to interface effectively with all the product assurance disciplines, as well as the engineering/manufacturing disciplines and management functions
3. Academic training/degree in engineering or science, supplemented by a good working knowledge of the application of quantitative methods (e.g., mathematics/statistics)
4. Working experience in at least one (preferably all) of the design assurance disciplines (e.g., reliability engineering), as well as manufacturing quality assurance
5. Strong desire to work as a generalist rather than to specialize in only one of the product assurance disciplines
6. A leaning toward a staff position, rather than a line management position
7. Good communication skills, both verbal and written

---

* It should be noted that the problem of finding generalists also exists within the individual disciplines (e.g., reliability)—people do tend to specialize even within the disciplines. Therefore, to a lesser degree, the following discussion also applies at the discipline level.

TABLE 14-3 Academic Training/Work Experience Combinations vs. Product Assurance Disciplines

	Work Experience							
Academic Training	Manufacturing	Design Engineering	Project Management	Test	Systems Engineering	Operations Research	Software Engineering	Industrial Engineering
Mechanical engineering	Q	R,M,S	R,M,S,Q	R,M,S,Q		R,M	R,M,A	R,M,Q
Systems engineering	Q	R,M,S	R,M	R,M,S,Q		R,M,S	R,M	R,M,Q
Physics	Q	R,M,S	R,M,S	R,M,S	R,M,S		A	
Mathematics/statistics		R,M,Q	R,M	Q	R,M	R,M,A	A	
Psychology		H		H	H			
Electronic/electrical engineering	Q	R,M,S	R,M,S,Q	R,M,S,Q	R,M,S	R,M,S	R,M,A	R,M,S,Q
Manufacturing engineering	Q	Q	Q	Q	Q			Q
Engineering management	Q	R,M,S		Q	R,M,S	R,M	M	R,M,Q
Computer science/engineering		R,M,A	A	A	R,M,A	R,M,A	A	R,M,Q

Legend:

R - Reliability
M - Maintainability
H - Human Factors
Q - Manufacturing Quality Assurance
S - Safety
A - Software Quality Assurance

246

Based on the diversity of the above qualifications, this type of individual is obviously rare in the product assurance field. When available, however, this person is extremely valuable to achieving the desired goals.

14.3 Employment Possibilities

At one time, employment possibilities in product assurance were rather limited—primarily in manufacturing quality assurance work and in military-oriented products. Just as the disciplines of product assurance have grown and come of age, so too have career opportunities in government, industry, consulting, and education. As discussed earlier in this chapter, the enormous technological advances in all fields have had the effect of enormously expanding assurance-related employment opportunities. As technology creates new fields, assurance-related activities are also created. Today, employment possibilities in product assurance abound.

14.3.1 Government

All of the major Department of Defense services (Army, Navy, and Air Force) employ product assurance specialists. Services such as the National Aeronautics and Space Administration (NASA) and the Department of Transportation (DOT) also employ these specialists. In a product assurance sense, the role of these specialists is to establish requirements to be imposed contractually, evaluate bidders' proposals, provide technical direction to contractors, monitor contractor performance against requirements, and perform independent technical analysis evaluation. They make sure that the government gets its money's worth relative to product integrity. To help these government agencies, there is the Defense Contracting Agency Services (DCAS), which provides centralized on-site (e.g., at contractor's facilities) product assurance monitoring support. DCAS representatives perform such functions as quality audits, decision-making relative to discrepant material, monitoring reliability and maintainability demonstration testing, and government buy-off at final quality inspection and test. Falling in the special-functions category are organizations such as the Reliability Analysis Center (RAC) and the Data & Analysis Center for Software (DACS), which are operated for the Air Force's Rome Air Development Center (RADC) by ITT Research Institute and which act as repositories of microelectronic and discrete semiconductor reliability information and software data tools and techniques information, respectively.

14.3.2 Industry

Product assurance job opportunities in industry are extensive, both in small and large companies, and in companies involved in both military and commercial products. In the small companies, product assurance job opportunities may sometimes be limited to those in manufacturing quality assurance; at the other extreme, however, a small company may be faced with more than quality requirements, thus necessitating individuals skilled in the various disciplines or a product assurance generalist. The latter individual would be more apt to be the choice in the case of tight budgets. In the large companies such as GTE Sylvania, General Electric, and Boeing, it is not uncommon to have specialists represented in all the product assurance disciplines. Depending on the type of product assurance organizational structure (see Chapter 4), these specialists assume the roles of reliability/maintainability project engineers, quality control engineers and analysts, inspection personnel, and maintenance engineers. When one considers all engineering, management, technical, and administrative personnel involved in product assurance work in typical large corporations, the number of product assurance personnel easily approaches 10 to 15 percent of the total work force, and the demand appears to be increasing.

14.3.3 Consulting

The consulting field also offers job opportunities in product assurance, although to a lesser degree than government and industry. Organizations in the consulting field can essentially be divided into two categories—for-profit and nonprofit organizations. In the former category are companies, such as ARINC Research Corporation, Planning Research Corporation, and Assurance Technology Corporation, which provide a wide variety of product assurance consulting services to both private industry and the government. In the latter category are companies, such as the MITRE Corporation, Battelle Memorial Institute, Rand Corporation, which provide similar services to government agencies (e.g., the Federal Aviation Agency and the Air Force). In addition, there are the quasi-consulting type of organizations which furnish product assurance–oriented personnel on a temporary basis to private industry to ease peak workload periods. Such organizations, which provide so-called "job-shoppers," are an important source of supply of product assurance personnel. Many times these job-shoppers end up going on direct employment with the company with whom they are job-shopping if it is to their mutual benefit.

14.3.4 Education

Finally, there are the product assurance job opportunities in education. Although somewhat limited, this is still an area to be reckoned with, both now and in the future, when the demand for product assurance specialists will intensify. More and more, the need for academically qualified product assurance personnel will increase, with a corresponding need for instructors to teach them. Earlier in this chapter, various sources of academic training for product assurance were described. These sources require instructors for part-time and full-time courses in the product assurance disciplines. Instructors for the part-time courses, which are aimed at upgrading and keeping an individual's expertise current, are usually drawn from industry. These instructors are usually eminently qualified persons from the product assurance disciplines who impart their knowledge to others from government and industry. Instructors for the full-time courses, which are given as part of a curriculum leading to a degree or certificate, are persons who have either expanded their horizons in the educational field (e.g., electronic engineering instructors who include courses related to reliability engineering as part of their repertoire), or have come to education from industry and set up courses (e.g., software quality assurance) in an attempt to bring the practical product assurance approach to the classroom.

14.4 Rewards

Taking a rather myopic view, one could think of rewards solely in terms of money. Certainly, this is one side of the coin. A career in product assurance *is* financially rewarding. Salaries of product assurance practitioners are certainly competitive with those of individuals in comparable disciplines. Salary surveys consistently show that salaries of individuals in the design assurance disciplines are unquestionably comparable with those of hardware design/software engineering personnel. Similarly, manufacturing quality assurance personnel are paid salaries which equal, if not exceed, those of manufacturing personnel.

Although this chapter deals primarily with the product assurance professional, there are many assurance career opportunities for the nonprofessional. Electronic and mechanical technicians are in increasing demand as the scope and depth of assurance activities increase. Special clerical and administrative opportunities have developed along with the assurance sciences. Highly skilled inspection and test personnel are now routinely required for the specialized activities of product verification, evaluation, calibration, and demonstration. Supervisory and managerial employment opportunities associated with these nonprofessional jobs have likewise increased.

Also, as pointed out earlier in this chapter, many of the nonprofessional jobs in the assurance field offer a growth path for hourly and weekly paid personnel. Inspection jobs are generally considered part of a higher skill level than production jobs and, therefore, are more highly paid. A technician, having developed special skills in test equipment calibration, for example, is generally more highly paid than a technician using the test equipment. Similarly, a quality control auditor, possibly moved in from a production/manufacturing job, will earn more in this more highly skilled position.

However, when contemplating any career, one must also think of perhaps a more important factor—job satisfaction. If you measure job satisfaction in terms of challenge, then a career in product assurance is for you! Where else can you be faced with the challenge of taking someone's "paper" design and making sure that it is translated into a piece of hardware which functions reliably, is easy to maintain, is safe to use, has aesthetic appeal, and is quality made? Considering the constraints (e.g., cost) that usually exist, this is no easy task. Overriding such constraints are the typical organizational pressures ("Quality be damned; we've got to meet our production quota for the month!") which are brought to bear, and which make the life of the product assurance practitioner interesting, to say the least. Besides being financially rewarding, a career in product assurance offers great challenge.

The nonprofessional product assurance practitioner also has considerable noncompensation rewards in the assurance fields. There is the satisfaction of an inspector signing approval, for his or her company, for the delivery of a quality product. The ability to resist production schedule pressures, while assuring product integrity by selectively stopping rejectable products, is a task that yields considerable satisfaction.

In addition to job satisfaction, the product assurance professional has many opportunities to achieve recognition in the various disciplines. The rapid technical development of all of the disciplines has resulted in a plethora of textbooks, magazines, symposium technical papers, and professional journal articles. These forums allow for the publishing and presentation of technical achievements in all of the assurance fields. Along with this type of recognition comes the possibility of being a member of one of the assurance professional societies such as the IEEE—Professional Group on Reliability or the American Society for Quality Control (ASQC). Local branch organizations exist from these professional societies that afford the opportunity to become officers and affect professional development. Associations such as the National Security Industrial Association (NSIA) and Electronic Industries Association (EIA) allow for company representation in the assurance disciplines. These committees form the interface between industry and the government. They allow for national and international recognition in all of the assurance disciplines. The professional societies also make it possible for an active contributing member to become recognized through membership upgrading as activities and contributions become more significant. Some of the societies even have professional certification, requiring minimum skill levels and contributions (e.g., ASQC). In fact, the assurance sciences seem to have the potential for more professional recognition than most technical professions as measured by publications, seminars, and symposia.

14.5 Prognosis

The prognosis for someone contemplating a career in product assurance is, without question, excellent. From modest beginnings covering product quality concern, the field of product assurance now embraces a variety of disciplines aimed at assuring product integrity. The fire of product assurance has been fanned so that its glow is now felt in almost every product line, from the quality-controlled aspirin to the safety-designed nuclear power reactor. Consequently, this growth has resulted in the need for product assurance specialists. If anything, the demand for such individuals is exceeding supply. In the long term, this need can only increase. The demand for product integrity, once primarily restricted to military products, is now being translated to commercial products. The need for product assurance requirements and professionals will be commonplace on *all* products. To keep up with this demand, there will have to be additional sources of academic training, perhaps even starting at the high school level. In short, a career in product assurance offers an individual the opportunity for a growth position performing stimulating and challenging work, while at the same time yielding good financial rewards.

The Company Product Assurance Program Planning Model

A.1 Product Assurance Program Plan

A.2 Introduction

This planning model is organized in four main parts: introduction, product assurance systems, design assurance, and quality assurance. This introduction, the first main part, embraces the overall product assurance program. It consists mainly of charts and matrices which present a great deal of information in a small space. In fact, this section is practically sufficient, by itself, to describe the product assurance program to a product assurance veteran. The last three sections serve to "flesh out" the program with more specific task definitions.

The introduction to a product assurance program plan should begin with a statement of the extent to which the company is committed to assuring the integrity of *this product,* the characteristics of this product which essentially define its integrity, and any relative priorities among these characteristics. Requirements or goals for these characteristics should be stated quantitatively, to the extent possible. All this provides a beginning rationale for the selection of product assurance tasks which are appropriate to this product, their scheduling, and the assignment of responsibilities within the organization.

A.2.1 Management and organization

The organization within the company which will have primary responsibility for execution of the product assurance program is shown graph-

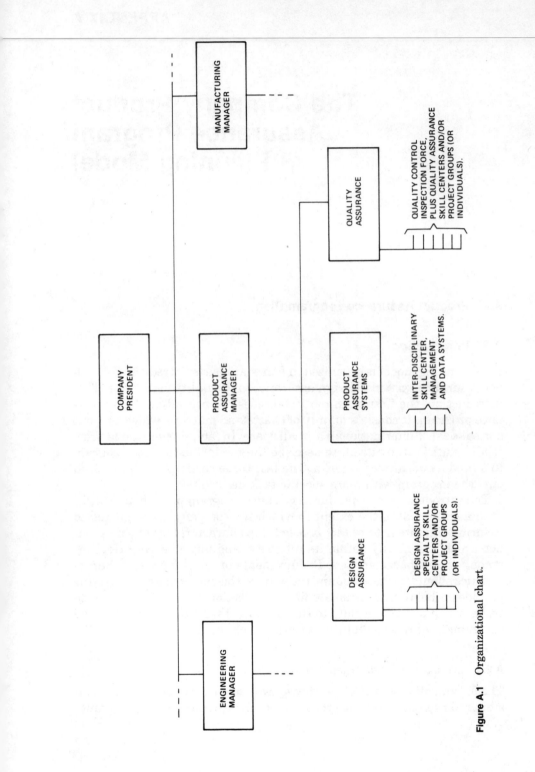

Figure A.1 Organizational chart.

ically on the organizational chart in Figure A.1. It is shown that the product assurance manager has direct access to top company management, direct lateral communication with the responsible design and production organizations, and direct line control of product assurance specialists at the working level. The product assurance organization is basically functional, but allows the assignment of product assurance project managers (or engineers) at any level—from product assurance task manager (out of product assurance systems), down to reliability project engineer (out of, say, the systems reliability group within design assurance).

A.2.2   Task selection

The specific product assurance tasks to be performed on this project are selected as those most appropriate to the unique requirements of this product and its intended application. Likewise, the degree of intensity, depth of analysis, degree of control, etc., appropriate to the various tasks is a function of product characteristics. The task matrix shown in Table A.1 gives a general idea of how task selection and emphasis is planned. For purposes of illustration, three characteristics (product use criticality, complexity, and production volume) are chosen and assigned only two degrees of existence each, so that there are only eight combinations of characteristics to deal with. Obviously, products and projects have other important characteristics which affect task selection, and these characteristics have almost infinite granularity, rather than mere two-step existences. Nevertheless, such a device is practical (and we believe, necessary) for specific industrial environments and products. Although an elaborate matrix would not be included in an actual program plan, a discussion of the reasons for major emphasis on certain tasks and minor emphasis on others is appropriate—particularly if the plan is to be reviewed and/or approved by the customer.

Notice that the task outline for this matrix has been considerably expanded, compared to the plan outline presented earlier. If this planning model were carried to its ultimate conclusion, the task outline would be even further expanded. The resolution or "granularity" of task emphasis or intensity would vary from task to task (not be limited to "low," "moderate," and "high"). Further, each level of intensity, for each task, would be clearly defined in succeeding sections of the planning model (each in a separately numbered subparagraph, perhaps—for reference by the task matrix). Again, such a planning model is practical for a limited range of product characteristics. For the general case considered here, the following sections of this planning model are relied upon to define the tasks generally and make reference to other sections of the book for further definition, including "what do low, moderate, and high mean?"

TABLE A.1 Task Selection Method

	Noncritical mission, benign environment				Critical mission, severe environment			
Product use criticality → Product complexity → Production volume	Simple		Complex		Simple		Complex	
Product assurance task	Low	High	Low	High	Low	High	Low	High
2.0 Product assurance systems								
2.1 Management systems								
2.1.1 Project management (hardware/software)	L	M	L	H	M	M	M	H
2.1.2 Configuration management (hardware/software)	L	L	H	H	M	M	H	H
2.1.3 Logistics management	L	L	M	M	L	L	H	H
2.1.4 Value engineering	L	M	H	H	L	M	L	H
2.1.5 Deliverable data management (hardware/software)	Mostly dependent on other factors							
2.2 Data systems (hardware/software)								
2.2.1 Product failure and discrepancy reporting and analysis	L	M	L	H	M	H	H	H
2.2.2 P.A. management data collection and analysis	Mostly dependent on other factors							
3.0 Design assurance								
3.1 Design guidelines								
3.1.1 Parts/materials selection and application	L	M	L	L	M	H	M	M
3.1.2 Electrical design	L	M	L	M	M	H	M	H
3.1.3 Physical design	L	L	M	M	H	H	H	H
3.1.4 Software design	L	M	L	M	M	H	M	H
3.2 Design analysis								
3.2.1 Systems effectiveness	L	L	M	M	M	M	H	H
3.2.2 Reliability/maintainability/systems safety	L	L	L	L	H	H	H	H
3.2.3 Environmental	L	L	L	L	H	H	H	H
3.2.4 Human factors/operability	L	L	M	M	M	M	H	H
3.2.5 Software features (e.g., robustness, portability)	L	L	M	M	M	M	H	H
3.3 Evaluation testing								
3.3.1 Environmental	L	M	L	M	H	H	H	H

Task						
3.3.2 Longevity/reliability	L	M	L	H	M	M
3.3.3 Maintainability/operability	L	L	M	M	H	H
3.3.4 Software validation/verification	L	M	L	H	M	M
3.4 Design review (hardware/software)						
3.4.1 Concept	L	L	H	M	H	H
3.4.2 Intermediate	L	L	H	M	H	H
3.4.3 Final	L	M	M	H	H	H
4.0 Quality assurance						
4.1 Initial preparations						
4.1.1 Quality/producibility design review	L	M	H	M	M	H
4.1.2 Work instructions	L	H	H	H	M	H
4.1.3 Records, corrective action, and quality costs	L	H	H	H	M	H
4.2 Facilities and standards						
4.2.1 Documentation and changes	L	M	H	M	M	H
4.2.2 Measurement and test	Mostly dependent on other factors					
4.3 Supplier control						
4.3.1 Responsibility	Full responsibility for supplier quality					
4.3.2 Incoming materials control	L	M	M	M	M	H
4.3.3 Corrective action	L	M	H	H	H	H
4.4 Manufacturing control						
4.4.1 Processing and fabrication	L	M	M	H	M	H
4.4.2 Final inspection and test	L	M	M	H	H	H
4.4.3 Handling, storage, and delivery	Mostly dependent on other factors					
4.4.5 Statistical quality control and analysis	H	H	H	H	L	H
4.4.6 Inspection status	L	L	H	M	L	H
4.5 Customer liaison	Mostly dependent on customer					
4.5.1 Customer participation	Requirements and company operating methods					
4.5.2 Customer property	Requirements and company operating methods					

*Key to relative task importance: L = low, M = moderate, H = high.

A.2.3 Task responsibilities

Overall management planning, scheduling, and implementation responsibility for the product assurance program reside with the product assurance organization, as described in the first section. In order to assure the technically proper and timely completion of the specific tasks cited in the preceding section, more specific assignments of task responsibilities are necessary. A summary of task responsibilities is provided in Table A.2.

Organizations cited as "support" on a given task are required to support the efforts of the "prime" organization for that task. The prime organization is held responsible to product assurance management for technically proper and timely task completion. On a large or tightly controlled program, the prime organization might write (and negotiate) detailed support task descriptions.

A.2.4 Task schedules

The performance of product assurance tasks on this program is scheduled in a manner to promote effective achievement of the product assurance program goals at minimum cost. They are scheduled so that each task is begun at the earliest time at which prerequisite data and authority are available and completed by the time the task output data are required as input for other tasks—or to support the decisions necessary to a major program event.

Figure A.2 provides an overview of the product assurance program. Task schedules are displayed in relation to each other and in relation to major program events. The major program events selected are:

1. Receipt of request for proposal (RFP)
   a. Receipt of request for quote, invitation to bid, etc.
   b. Favorable market survey report
   c. In general, any event that precipitates the serious consideration to develop and/or manufacture a new product
2. Proposal submittal to the customer, in the case of 1.a above, and to top company management in any other case
3. Contract award, or development go-ahead from top company management
4. Fabrication of hardware development models (units)
5. Drawing release or, in general, authorization to proceed with pilot production, followed by product qualification or, in general, prove out of producibility and authorization to proceed with mass production
6. Release for software distribution
7. Building and delivery of the end product

TABLE A.2 Product Assurance Task Requirements

MAJOR PRODUCT ASSURANCE TASKS	ORGANIZATIONAL UNITS				
		PRODUCT ASSURANCE			
	DESIGN ENGINEERING	DESIGN ASSURANCE	PRODUCT ASSURANCE SYSTEMS	QUALITY ASSURANCE	MANU-FACTUR-ING
2.0 PRODUCT ASSURANCE SYSTEMS					
2.1 MANAGEMENT SYSTEMS		S	P	S	
2.2 DATA SYSTEMS		S	P	S	
3.0 DESIGN ASSURANCE					
3.1 DESIGN GUIDELINES	S	P	S		
3.2 DESIGN ANALYSIS	S	P	S		
3.3 EVALUATION TESTING	S	P	S	S	
3.4 DESIGN REVIEW	S	S	P	S	S
4.0 QUALITY ASSURANCE					
4.1 INITIAL PREPARATIONS			S	P	S
4.2 FACILITIES AND STANDARDS				P	S
4.3 SUPPLIER CONTROL		S	S	P	S
4.4 MANUFACTURING CONTROL			S	P	S
4.5 CUSTOMER LIAISON	S	S	P	S	S

Key: P = Prime Responsibility
S = Support Responsibility

259

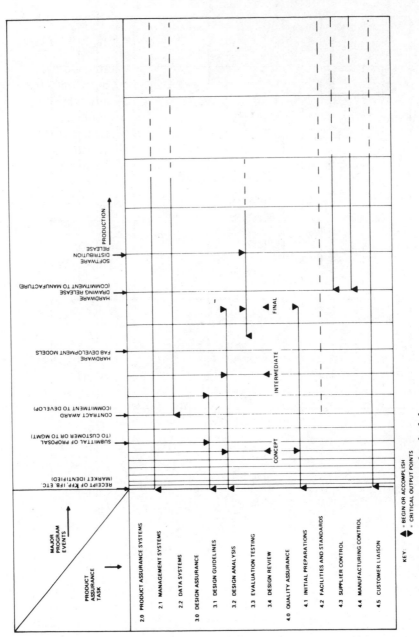

Figure A.2 Product assurance task schedule.

260

These are all key events to which *all* program tasks, including the product assurance tasks, are keyed to achieve.

A.2.5 Documentation

There are three basic classes of documentation associated with the product assurance program: (1) requirements documents, which describe *what* must be done and, to varying degrees, how, when, where, and by whom; (2) reference documents, which provide more detailed advice, generally concerned with how to perform the tasks in such a way as to satisfy the requirements; (3) product assurance program documentation consisting of the records and reports of accomplishments.

An important aspect of requirements documents is their order of precedence, which is—generally:

1. The customer's product specification
2. The customer's statement of work, or equivalent
3. The company's product specification
4. The company's overall program plan
5. The product assurance program plan

Once these are satisfied, any reference documents cited by them form the next level of authority for task planning and performance. Some typical documentation relationships are shown by Table A.3.

The product assurance program documentation provides evidence of accomplishment and historical data. The usual view taken by the customer and by company management is—*if it isn't documented, it isn't done*—and rightly so.

A.2.6 Product assurance systems

Certain mechanisms, which are called product assurance systems, serve to help bridge the communication gaps between the product assurance specialty groups; between these groups and design, or manufacturing; and between product design, manufacture, and field support phases. These systems, not being assurance sciences, are not described in any depth in this book. They are adequately described in other publications, and they must be considered in product assurance program planning. Such systems are grouped into two classes: (1) product assurance management systems and (2) product assurance data systems.

TABLE A.3 Documentation Associated with the Product Assurance Program

Product assurance program	This model, sub-section	This book, chapter	Reference Documents	Records & reports from product assurance program
			Government documents (important examples)	
Management and organization	A.2.1	4, 6, 13	MIL-STDS-499, 470, 785, 882, 790, 1679, 1472, MIL-S-52779	This plan, company policy documents, organization charts
Task selection	A.2.2	5	Statements of work, referencing above.	This plan, company policies and procedures
Task responsibilities	A.2.3	5	N/A	This plan, company policies and procedures
Task schedules	A.2.4	5, 6	Statements of work	This plan, progress reports
Documentation	A.2.5	7, 13	Various data item descriptions, (DID) DD 1664	This plan, company policies and procedures
Product assurance systems	A.2.6	5		This plan, company policies and procedures
Management systems	A.2.7	4, 6, 8		Periodic reports
Data systems	A.2.8	8		Periodic reports
Design assurance	A.2.9	2, 13	MIL-STDS-470, 785, 882, 1679	This plan, company policies and procedures
Design guidelines	A.2.10	8, 13	NAVSHIPS 94501, AFSC DH-9, HI-3 MIL-STD-1679	Published guidelines
Design analysis	A.2.11	9	MIL-HDBK-217, 472, 1679	Published reports
Evaluation testing	A.2.12	11, 12, 13	MIL-STD-471, 781, 1679	Test plans and reports
Design review	A.2.13	8, 13	MIL-STDS-470, 785, 882, 1679	Minutes of meetings

Quality assurance	A.3	2, 13	MIL-Q-9858, MIL-S-52779, MIL-I-45208	This plan, company policies and procedures
Initial preparations	A.3.1	8	MIL-Q-9858	Detailed plans, minutes of meetings
Facilities and standards	A.3.2	10	MIL-Q-9858, MIL-C-45662	Company procedures, records
Supplier control	A.3.3	10, 13	MIL-Q-9858, MIL-STD-105, 414, MIL-H-106, 107, 108	Procedures, inspection and test records
Manufacturing control	A.3.4	10, 11, 12	MIL-Q-9858, MIL-STD-105, 414, MIL-H-106, 107, 108	Procedures, inspection and test records
Customer liaison	A.3.5	12	MIL-Q-9858	Procedures, records

A.2.7   Product assurance management systems

The most common product assurance management system is product assurance project management, which "projectizes" the product assurance effort on a given project, even in the presence of a basically function-oriented product assurance organization. It is established by delegating overall product assurance responsibility (and the necessary authority) on a given project to an individual or group. This promotes project-oriented communication across functional boundaries within the company.

Product assurance management systems also include configuration management, logistics management (or integrated logistics support—ILS), value engineering, and data management. These systems are all concerned with achieving some aspect of product integrity in the most efficient manner. They may be implemented by project (i.e., as parts of product assurance project management) or by function.

A.2.8   Product assurance data systems

The collection, analysis, and reporting of product performance, failure, software error, maintenance, quality discrepancy, and customer complaint data forms a large and important part of the product assurance activity in a manufacturing company. Systems designed to perform this task, whether manual or automated, are necessary "loop-closers" for the various product assurance specialties and for the product assurance function.

Similar systems, whether unique to the product assurance organization or common to company management, are necessary to collect, analyze, and report work performance, cost, and schedule data.

A.2.9   Design assurance

The sole mission of design engineering is to produce the documentation necessary to the creation of the hardware or software product. It is the mission of design assurance to ensure that this occurs in a comprehensive manner and that the result provides designed-in product integrity commensurate with cost and customer requirements. In the case of software, this mission is entrusted to organizations which are given the name "software quality assurance." Nevertheless, the general task areas are the same, whether they be aimed at hardware or software. All projects which result in the design (and not necessarily fabrication) of hardware or software should have the conscious application of design assurance guidance, analysis, evaluation, and review.

A.2.10 Design guidelines

Before detailed design begins, more or less broad design guidelines should be established for:

The selection and application of parts and materials

Electrical design (grounding philosophy, test points, etc.)

Physical design (packaging philosophy, environment, etc.)

Software design (data structures, program design language, etc.)

A.2.11 Design analysis

Throughout the design process, the evolving design must be analytically examined for adherence to product integrity requirements—to uncover and accomplish trade-offs to enhance product integrity characteristics. Such analysis should include, as appropriate: systems effectiveness, hardware/software reliability and maintainability, personnel and equipment hazard, failure modes and effects, producibility, electromagnetic/radio frequency susceptibility/radiation, mechanical/ thermal stress, and human factors/operability analyses, software robustness and portability—to name a few.

A.2.12 Evaluation testing

As soon as suitable hardware or software is available, those product integrity characteristics that (1) are critical to this product or (2) are indicated by analysis to have a high risk of not being achieved or (3) cannot be analytically determined with a satisfactory degree of confidence should be evaluated by test. "Suitable hardware or software" means hardware or software which can be reasonably expected to represent the end product—at least insofar as the characteristic under evaluation is concerned.

A.2.13 Design review

The objective of formal design review is to assure that the hardware and software design process is systematic, scientific, and comprehensive. On a given project, design reviews seek to assure that *all* specified hardware and software design objectives are met, that selected designs are the optimum solutions to design problems, and that the final design is adequately documented and economically producible. There are three general classes of design reviews: concept, intermediate, and final. More than one of each class may be held on a project.

Design reviews use the results of preceding design assurance activities as input data and provide guidance for succeeding design assurance, and quality assurance, activities.

The *concept* design review assures that all necessary product performance requirements are established, that feasible design approaches are available, that external interfaces are defined, and that areas of risk are identified.

The *intermediate* design review seeks to evaluate the areas of risk, select a preferred design approach, define internal interfaces, and assure that physical realization is feasible.

The *final* design review assures that all significant risks have been evaluated and found acceptable, that parts and materials will be available for manufacture, that design objectives have been met, that the design is properly documented, and that it is economically producible.

A.3   Quality Assurance

The manufacturing function must fabricate products which are faithful to the documented design. Quality assurance is charged with assuring that the durable quality and integrity intended by the design is consistently realized in all copies of the product. It includes both the acceptance function (e.g., incoming and process inspection) and the prevention function (e.g., process control, experimental design). Quality assurance is concerned with deliverable software and documentation quality, as well as product hardware.

A.3.1   Initial preparations

It is imperative that potential problems in achieving, verifying, and recording evidence of product quality be identified and prepared for as early as possible. Quality assurance should review the contract to identify and provide for any special or unusual quality requirements. Advance planning for quality assurance task workload, facilities, and equipment availability, and any special skill requirements must be accomplished in concert with the manufacturing plan. Quality assurance must be cognizant of design developments which could affect the difficulty of achieving or verifying product quality. To that end, quality assurance representatives should participate in the appropriate design reviews in order to: (1) identify any peculiar skill, equipment, or facility requirements made necessary by the design and (2) influence the design to accommodate manufacturing processes which will promote ease of achievement of consistent quality. Work and inspection instructions must be prepared.

A.3.2   Facilities and standards

Quality assurance must have access to gauges, instruments, and other measuring and testing equipment necessary to ensure that supplies conform to specified requirements. Such instruments must be correlated with standards, the accuracy of which are traceable to the National Bureau of Standards. An effective documentation and configuration control system must exist.

A.3.3 Supplier control

The company is fully responsible for the quality of all component parts and materials supplied by vendors, which constitute (or otherwise affect the quality of) the product (including deliverable software, data, and documentation). Quality assurance must, therefore, have a system of supplier material inspection and corrective action. In cases where the customer has a voice in controlling supplier material and data, that voice must be effectively transmitted to the suppliers by quality assurance.

A.3.4 Manufacturing control

Inspection stations must be maintained at appropriately located points in the manufacturing operations to assure continuous control of quality of parts, components, and assemblies. Process-type inspection should be provided when direct inspection of hardware is inadequate or inadvisable. Final inspection of the end product for conformance to requirements must be accomplished. Control and audit of product and component handling, storage, and delivery is required. Nonconforming material must be segregated and its disposition controlled and recorded. Scrap and rework costs data should be collected, analyzed, and used to promote optimization of manufacturing and quality control systems. Appropriate statistical quality control and analysis techniques should be used to improve the effectiveness and reduce the costs of quality control.

A.3.5 Customer liaison

Normally, there are a number of coordinated customer/company actions which must take place, with respect to the production and support of the product. The extent of the customer's rights of access to manufacturing facilities, process data, quality records, etc., must be defined and agreed upon. Any customer access to the company's suppliers or rights to on-premises inspection must be planned and provided for. In the event that the product comprises any customer-furnished equipment, the need to provide for meticulous record-keeping of the receipt handling, modification, storage, etc., of such equipment is obvious.

Bibliography

B.1 General References

B.1.1 Reliability

Amstadter, B. L. *Reliability Mathematics: Fundamentals, Practices, Procedures.* New York: McGraw-Hill, 1971.

Arsenault, J. E., and J. A. Roberts, eds. *Reliability & Maintainability of Electronic Systems,* Potomac, Md.: Computer Science Press, 1980.

Barlow, R. E., and F. Proschan. *Mathematical Theory of Reliability.* New York: Wiley, 1965.

Barlow, R. E., and F. Proschan. *Statistical Theory of Reliability and Life Testing.* New York: Holt, Rinehart, and Winston, 1974.

Basovsky, I. *Reliability Theory and Practice.* Englewood Cliffs, N.J.: Prentice-Hall, 1961.

Bompas-Smith, J. H. *Mechanical Survival: The Use of Reliability Data.* London, New York: McGraw-Hill, 1973 (edited by R. H. W. Brook).

Bourne, A. J., and A. E. Greene. *Reliability Technology.* New York: Wiley-Interscience, 1972.

Brook, R. H. W. *Reliability Concepts in Engineering Manufacture.* New York: Wiley, 1972.

Calabro, S. R. *Reliability Principles and Practices.* New York: McGraw-Hill, 1962.

Carter, A. C. S., *Mechanical Reliability.* New York: Halsted Press, 1973.

Dhillon, B. S., and C. Singh. *Engineering Reliability, New Techniques and Applications.* New York: Wiley-Interscience, 1981.

Dummer, G. W. A. *An Elementary Guide to Reliability.* Oxford, New York: Pergamon Press, 1974.

Enrick, N. L. *Quality Control and Reliability.* New York: Industrial Press, 1972.

Glass, R. *Software Reliability Guidebook.* Englewood Cliffs, N.J.: Prentice-Hall, 1979.

Gryna, F. M., Jr., eds. *Reliability Training Text.* New York: Institute of Radio Engineers, 1960.

Haviland, R. P. *Engineering Reliability and Long Life Design.* Princeton, N.J.: Van Nostrand, 1964.

Henley, E. J., and H. Kumamoto. *Reliability Engineering and Risk Assessment.* Englewood Cliffs, N.J.: Prentice-Hall, 1981.

Ireson, W. G. *Reliability Handbook.* New York: McGraw-Hill, 1966.

Jardine, A. K. S. *Maintenance Replacement and Reliability.* New York: Wiley, 1973.

Kivenson, G. *Durability and Reliability in Engineering Design.* New York: Hayden, 1971.

Kopetz, H. *Software Reliability.* London: McMillan Press, 1979.

Landers, R. R. *Reliability and Product Assurance: A Manual for Engineering and Management.* Englewood Cliffs, N.J.: Prentice-Hall, 1963.

Leake, C. E. *A Simplified Presentation for Understanding Reliability.* Pasadena, Calif.: Pasadena Lithographers, 1960.

Lloyd, D. K., and M. Lipow. *Reliability: Management, Methods, and Mathematics.* Englewood Cliffs, N.J.: Prentice-Hall, 1962.

Locks, M. O. *Reliability, Maintainability, and Availability.* New York: Spartan Books, 1973.

Longbottom, R. *Computer System Reliability,* New York: Wiley, 1980.

Myers, G. J. *Software Reliability through Composite Design.* New York: Petrocelli Books, 1975.

Myers, G. J. *Software Reliability Principles and Practice.* New York: Wiley-Interscience, 1976.

O'Connor, P. D. T. *Practical Reliability Engineering.* Philadelphia: Hayden, 1981.

Pieruschka, E. *Principles of Reliability.* Englewood Cliffs, N.J.: Prentice-Hall, 1963.

Roberts, N. H. *Mathematical Methods in Reliability Engineering.* New York: McGraw-Hill, 1964.

Sandler, G. H. *System Reliability Engineering.* Englewood Cliffs, N.J.: Prentice-Hall, 1963.

Shooman, M. L. *Probabilistic Reliability: An Engineering Approach.* New York: McGraw-Hill, 1968.

Sieworek, D. P., and R. S. Swarz. *The Theory and Practice of Reliable System Design.* Burlington, Mass.: Digital Press, 1974.

Smith, D. J. *Reliability Engineering.* New York: Barnes & Noble Books, 1972.

Smith, O. *Introduction to Reliability in Design.* New York: McGraw-Hill, 1975.

Von Alven, W. H., ed. *Reliability Engineering.* Englewood Cliffs, N.J.: ARINC Research Corporation, 1964.

B.1.2 Maintainability

Ankenbrandt, F. L. *Maintainability Design.* Elizabeth, N.J.: A. C. Books, 1963.

Blanchard, B. S. *Logistics Engineering and Management.* Englewood Cliffs, N.J.: Prentice-Hall, 1974.

Blanchard, B. S., and E. E. Lowery. *Maintainability: Principles and Practices.* New York: McGraw-Hill, 1969.

Cunningham, C. E., and W. Cox. *Applied Maintainability Engineering.* New York: Wiley, 1972.

Goldman, A. S., and T. B. Slattery. *Maintainability: A Major Element of System Effectiveness.* New York: Wiley, 1964.

Gordon, F. *Maintainability Engineering: Organization and Management.* New York: Halsted Press-Wiley, 1973.

Jardine, A. K. S. *Maintenance Replacement and Reliability.* New York: Halsted Press-Wiley, 1973.

Lientz, B. P., and E. B. Swanson. *Software Maintenance.* Stoneham, Mass.: Addison-Wesley, 1980.

Ross, M. A. *Designing for Minimal Maintenance Expense.* New York: Dekker, 1985.

Smith, D. J., and A. H. Babb. *Maintainability Engineering.* New York: Wiley, 1973.

B.1.3 Human factors

Chaney, F. B., and C. B. Harris. *Human Factors in Quality Assurance.* New York: Wiley, 1969.

Chapanis, A. R. E. *Research Techniques in Human Engineering.* Baltimore: John Hopkins Press, 1959.

Croney, J. *Anthropometrics for Designers.* New York: Van Nostrand Reinhold, 1971.

Fogel, L. J. *Biotechnology: Concepts and Applications.* Englewood Cliffs, N.J.: Prentice-Hall, 1963.

Gagne, R. M. *Psychological Principles in System Development.* New York: Holt, Rinehart and Winston, 1962.

Harris, D. H., and F. B. Chaney. *Human Factors in Quality Assurance.* New York: Wiley, 1969.

McCormick, E. J. *Human Factors Engineering.* New York: McGraw-Hill, 1970.

Meister, D., and G. F. Rabideau. *Human Factors Evaluation in Systems Development.* New York: Wiley, 1965.

Roebuck, J. A. *Engineering Anthropometry Methods.* New York: Wiley-Interscience, 1975.

Sell, R. C. *Human Factors in Work Design and Production*. New York: Halsted Press, 1977.

Woodson, W. E. *Human Engineering Guide for Equipment Designers*. Berkeley: University of California Press, 1954.

B.1.4 Quality assurance

Bersoff, E., et al., *Software Configuration Management: An Investment in Product Integrity*. Englewood Cliffs, N.J.: Prentice-Hall, 1980.

Burr, I. W. *Engineering Statistics and Quality Control*. New York: McGraw-Hill, 1953.

Burr, I. W. *Elementary Statistical Quality Control*. New York: Dekker, 1979.

Carter, C. L. *The Control and Assurance of Quality*. Dallas: Taylor Publishing Co., 1968.

Charbonneau, H. C. *Industrial Quality Control*. Englewood Cliffs, N.J.: Prentice-Hall, 1978.

Cho, C. *An Introduction to Software Quality Control*. New York: Wiley, 1980.

Chorafas, D. N. *Statistical Processes and Reliability Engineering*. Princeton, N.J.: Van Nostrand, 1960.

Cowden, D. J. *Statistical Methods in Quality Control*. Englewood Cliffs, N.J.: Prentice-Hall, 1957.

Crosby, P. B. *Cutting the Cost of Quality: The Defect Prevention Workbook for Managers*. Boston: Industrial Education Institute, 1967.

Crosby, P. B. *Quality Is Free: The Art of Making Quality Certain*. New York: McGraw-Hill, 1979.

Dewar, D. I. *The Quality Circle Guide to Participative Management*. Englewood Cliffs, N.J.: Prentice-Hall, 1982.

Duncan, A. J. *Quality Control and Industrial Statistics*. Homewood, Ill.: Richard D. Irwin, 1965.

Dunn, R., and R. Ullman. *Quality Assurance for Computer Software*. New York: McGraw-Hill, 1981.

Feigenbaum, A. V. *Total Quality Control: Engineering and Management*. New York: McGraw-Hill, 1961.

Grant, E. L., and R. S. Leavenworth. *Statistical Quality Control*. New York: McGraw-Hill, 1972.

Gryna, F. M., and J. B. Juran. *Quality Planning and Analysis*. New York: McGraw-Hill, 1970.

Juran, J. M. *Quality Control Handbook*. New York: McGraw-Hill, 1962.

Myers, G. F. *The Art of Software Testing*. New York: Wiley-Interscience, 1979.

Nixon, F. *Managing Costs to Achieve Quality and Reliability*. New York: McGraw-Hill, 1971.

Romig, H. G., and G. E. Hayes. *Modern Quality Control*. Encino, Calif.: Benziger, Bruce, and Glencoe, 1977.

Ross, J. E. *Japanese Quality Circles and Productivity*. Reston, Va.: Reston Publishing, 1982.

Shewhart, W. A. *Economic Control of Quality of Manufactured Product*. New York: Van Nostrand, 1931.

B.1.5 Safety

Hammer, W. *Handbook of System and Product Safety*. Englewood Cliffs, N.J.: Prentice-Hall, 1972.

Kolb, J., and S. Ross. *Product Safety and Liability*. New York: McGraw-Hill, 1980.

Malasky, S. W. *System Safety, Planning Engineering Management*. Rochelle Park, N.J.: Hayden Books, 1974.

Peters, G. A. *Product Liability and Safety*. Washington, D.C.: Coiner Publications, Ltd., 1971.

Peterson, D. *Techniques of Safety Management*. New York: McGraw-Hill, 1978.

B.2   Military

B.2.1   General

MIL-STD-721, " Definitions of Effectiveness Terms for Reliability, Maintainability, Human Factors, and Safety."

B.2.2   Reliability

MIL-STD-785, "Reliability Program for Systems and Equipment Development and Production."
MIL-STD-756, "Reliability Prediction."
MIL-HDBK-217, "Reliability Prediction of Electronic Equipment."
MIL-STD-781, "Reliability Tests: Exponential Distribution."
MIL-STD-1635, "Reliability Growth Testing."
MIL-HDBK-189, "Reliability Growth Handbook."

B.2.3   Maintainability

MIL-STD-470, "Maintainability Program Requirements (for Systems and Equipments)."
MIL-HDBK-472, "Maintainability Prediction."
MIL-STD-471, "Maintainability Demonstration."
DH1-9, AFSC Design Handbook, "Maintainability."

B.2.4   Human factors

MIL-STD-1472, "Human Engineering Design Criteria for Military Systems Equipment and Facilities."
MIL-H-46855, "Human Engineering Requirements for Military Systems, Equipment and Facilities."
DH1-3, AFSC Design Handbook, "Personnel Subsystems."

B.2.5   Quality assurance

MIL-Q-9858, "Quality Program Requirements."
MIL-P-11268, "Parts, Materials, and Processes Used in Electronic Equipment."
MIL-STD-105, "Sampling Procedures and Tables for Inspection by Attributes."
MIL-STD-414, "Sampling Procedures and Tables for Inspection by Variables for Percent Defective."
MIL-S-52779, "Software Quality Assurance Program Requirements."

B.2.6   Safety

MIL-STD-882, "System Safety Program for Systems and Associated Subsystems and Equipment: Requirements for" NAVORD OD 44949, "Weapon System Safety Guidelines Handbook."
DH1-6, AFSC Design Handbook, "System Safety."

Index

About the Authors

Eugene R. Carrubba has more than 25 years' experience in the product and quality assurance fields. Currently manager of product assurance for Symbolics, Inc., he has held positions with Digital Equipment Corporation, GTE, AVCO, RCA, and Raytheon in various technical and management capacities related to product assurance. A registered professional engineer, Mr. Carrubba is a member of IEEE and an instructor at Western New England College.

Ronald D. Gordon is currently responsible for management of all manufacturing, procurement, quality assurance and testing for the Communications Sytems Division of GTE. Previously with Northrup as reliability engineering supervisor, Sperry as reliability/test engineer, and General Motors, he is a graduate of General Motors Institute in mechanical engineering and an active member of IEEE.